"Fr. Georges DeSchrijver is one of only a handful of contemporary writers on religion and science who exhibits equal competence in the history of Western philosophy and theology and in natural science from ancient times to contemporary quantum theory. Most writers are experts in one area or the other but not both. Hence, Fr. DeSchrijver's defense of a Trinitarian process-oriented understanding of the relation between religion and science deserves careful reading. Moreover, he explains subtle arguments from both philosophy/theology and natural science in relatively straightforward terms suitable for the educated lay reader. The book is thus suitable for use in undergraduate classrooms."
—Joseph Bracken, SJ, is professor emeritus of Theology of Xavier University in Cincinnati, Ohio

"Written with the clear prose and deft illuminating touch of a master teacher, this book fills a real gap for theologians, other scholars, and lay readers alike. Its engaging trek through the history of thought about the cosmos delivers not only scientific information but also theological insight as to how each cosmology shapes our understanding of the God who creates. A valuable resource for grappling with how today's move from a static to an evolving picture of the world shapes our understanding of the divine."
—Dr. Elizabeth Johnson, CSJ, Distinguished Professor of Theology, Fordham University (New York), Past President, Catholic Theological Society of America

How we imagine and think about the universe has an enormous influence on the way we see its Creator. This book traces the cosmologies that have shaped our thought, and provided the context for our theologies of God, from Genesis and Plato to Stephen Hawking and beyond. A book of enormous learning, it is, nevertheless, wonderfully clear and accessible, honed by years of teaching. It will be of great help to teachers and students of theology, as well as general readers seeking to understand the universe and it relationship to God.
—Denis Edwards, Australian Catholic University

"With expert attention to the tangled histories of scientific and theological accounts of the nature of the universe, De Schrijver prepares here fertile ground for an image of the Creator God as One who endows the world with

its own 'intrinsic creativity.' *Imagining the Creator* provides a rich vision of contemporary Christian cosmology in which creation is always new."
—Julia Feder, Creighton University

In his book *Theology for a Scientific Age*, the late scientist theologian Arthur Peacocke stated that theology is "most creative and long-lasting when it has responded most positively to the challenges of its times." *Imagining the Creator God: From Antiquity to Astrophysics* by Georges De Schrijver, SJ, is exemplary in this regard, a *tour de force* of research and intellect that persuasively demonstrates how theology has responded to the emerging insights of both philosophy and science throughout the ages to articulate its understanding of the Creator God. With skill and creativity, Fr. Georges orchestrates the voices of an amazing choir of scholars, ancient and contemporary; demonstrates the resonance between their worldviews and prevailing images of God; and challenges both readers and theologians alike to move beyond static conceptions of God and creation ensconced in outdated cosmologies and modes of thought. The lenses of contemporary science enable us to see that God is revealing something new and Fr. Georges exhorts us to perceive it.
—Gloria L. Schaab, SSJ, PhD, Professor of Theology, Director of Graduate Programs in Theology and Ministry, Associate Dean for General Education, Barry University

I had the pleasure of having many conversations with Father Georges when he stayed with the Alemany Brothers at Saint Mary's College during his visits here, and was able to read this brilliant work in manuscript form as the individual chapters were produced and were being edited by Father Dave. It is great to see the final product of years of good, careful scholarly work and reflection. Father Georges' Imagining the Creator God: From Antiquity to Astrophysics is a tour de force in solid Catholic theology, a theology that values truth wherever it is found and that seeks to integrate truth from many disciplines and perspectives in the search for wisdom, and, in this case, for wisdom about the cosmos and about the God who created it. I highly recommend this superb book.
—Brother S. Dominic Ruegg, FSC, PhD, Professor Emeritus of Catholic Theology, Saint Mary's College of California

IMAGINING
THE CREATOR GOD

IMAGINING
THE CREATOR GOD
FROM ANTIQUITY TO ASTROPHYSICS

Georges De Schrijver, S.J.

WIPF & STOCK · Eugene, Oregon

Wipf and Stock Publishers
199 W 8th Ave, Suite 3
Eugene, OR 97401

Imagining the Creator God
From Antiquity to Astrophysics
By Schrijver, Georges De, S. J.
Copyright©2015 Ateneo De Manila University Press
ISBN 13: 978-1-5326-1016-5
Publication date 9/30/2016
Previously published by Ateneo De Manila University Press, 2015

Contents

Preface

Imagining the Creator God: From Antiquity to Astrophysics is the result of many years of teaching on this subject. The first sketch of it is to be found in a course entitled "Ancient and New Creation Narratives" which I taught in the academic year 1992–1993 at the theological faculty of the Catholic University in Leuven, Belgium, where I held the chair of Foundational Theology. Foundational Theology comprises many topics: the influence of modern and post-modern thought on Christian theologizing, the phenomenon of liberation theologies and interreligious dialogue, and the encounter with deism and atheism. It was the latter topic that occasioned me to engage in the study of classic philosophical and modern scientific cosmologies. I must have taught that course three or four times before my retirement from the Catholic University in October 2000.

Once I became an emeritus professor I was free to accept invitations to teach, speak, and write internationally. One such invitation came from the Ateneo de Manila University, where from the academic year 2002–2003 on I taught on an annual basis the course Cosmology (Philosophy of Nature) in the Department of Philosophy. The content and title of this course evolved over the years: "Cosmology from Plato to Einstein and the Big Bang," "Cosmology from Plato to String Theory," and finally "A Brief History of Cosmology."

In this panoramatic overview a great many names and systems come into view: due attention is given to the first creation story in Genesis, to Plato's creation narrative in the dialogue *Timaeus,* and to Aristotle's account of the heavenly machinery and the way it regulates the life cycles on Earth, as well as to Plotinus' system of cosmic emanations. This is followed by an examination of the influence of Greek cosmology on the Christian doctrine of creation. This influence is evident in the Nicene Creed and the role it attributes to Christ, the Logos, in the creation of the world; in St. Bonaventure's view of Christ, the expressive Word of the Father, from whom the theo-expressive character of the created realm derives; and in Aquinas' wrestling with the problem of the eternity of the heavenly machinery.

In this overall religious milieu, the role of the Creator God is imagined in various ways: in Hebrew thought God is presented as the sovereign master, who through the power of his voice brings about order in the world, whereas Greek cosmologies, in addition, honor the intrinsic creativity of the world. Plato attributes a preeminent role to the World Soul, who lures chaotic matter towards receiving the imprint of geometrical forms; Aristotle, in turn, postulates the existence of 55 cosmic intelligences that, in imitation of the highest God, seduce their cosmic sphere to perfectly performing its specific rotation. Even the Cosmic Christ is assumed to animate the whole creation from within.

Christian cosmology is a harmonization of Hebrew thought and Greek cosmology. The picture of the Cosmic God that results from it is still present in the liturgies of East and West. The majestic triune God is enthroned above the firmament, a firmament to which he imparts a rotation that, in varied ways, will be communicated to the transparent concentric rings on which the respective planets Saturn, Jupiter, Mars, Venus, Mercury, the Sun, and the Moon are attached—and which all owe their proper inclination to an angelic or cosmic intelligence (in Aristotle's parlance) who gives a particular imprint to their rotation. In this way divine energy is gradually descending down from the highest heaven to the Earth that lies in the center of the heavenly machinery. From God in the highest not only flows the kinetic force that regulates the earthly rhythms of growth in plants and animals, but also the special energy—termed *grace*—that animates and renews the souls of humans. It is in this immortal celestial domain, too, that hosts of angels—the invisible orders of sovereignties, dominions, and thrones—incessantly chant the glory of the triune God. With this chant of praise the Christian community joins in whenever in the Eucharist they conclude the solemn preface with the words: "And so, with Angels and Archangels, with thrones and dominions, and with all the hosts and powers of heaven we sing the hymn of your glory: Holy, Holy, Holy, Lord God of hosts. Heaven and earth are full of your glory. Hosanna in the highest."

The Copernican turn and its confirmation by Galileo and Kepler dealt a serious blow to this picture of the world. In his attempt to simplify the Ptolemaic system of cycles and epicycles, Copernicus realized how much simpler it would be if the earth and the planets were understood to be revolving around the Sun instead of, as previously held, the heavenly machinery orbiting the earth. As a result, a series of ancient cosmological dogmas had to give way. In the heliocentric system, no place is left for the neat separation Aristotle imagined to exist between the indestructible heavenly domain and the impermanent sublunary realm (with his elementary telescope Galileo observed that the moon landscape was pretty much the same as that of the Earth); nor can the classic Greek concepts of the perfect circularity and utmost regularity of the planetary motions be upheld (Kepler showed that the planets move in an ellipse with the sun at one of the two foci and that they accelerate and decelerate their speed in proportion to their proximity to or distance from the sun).

In short, the new scientific mind is born in search of the universal physical laws of nature testable by experiments. Newton figures as the apogee of this new mentality. His universal law of gravitation explains both why things fall on the earth and the seemingly permanent rhythm of the planetary motions. The force of gravity pierces the void and acts upon things at a distance. The planets perform their fixed orbits around the sun on the basis of mutual attraction, which results in the formula: $F = Mm/r^2$ (the force of gravity equals the mass of the sun times that of the planet divided by the square of their distance). For Newton, the rational and harmonious order of the universe is revealed to us in the language of mathematics. It is with the help of mathematics and thanks to his profound insight into the physical reality of things that he proved the inconsistency of Descartes' highly speculative vortex theory (according to which the planets retain their orbits about the sun through a wheel of whirling tiny matter that creates a counter-pressure to their supposed centripetal drift).

Newton's universe is definitely static, framed as it is within the coordinates of absolute space and absolute time which, so to speak, forms the medium and "organ" through which the Eternal One regulates the mechanics of the universe. Just as is the case with the Hebrew God, the Newtonian God imposes his sovereign power upon the world; He does so through his use of the force of gravity. This picture of the world and of the Creator basically remained in place for almost three centuries until the year 1905 when Albert Einstein with his Special Theory of Relativity relativized Newton's notions of absolute space and absolute time. For Einstein, there is only one absolute: the speed of light that is always measured as 300.000 km/sec no matter the speed of the source emitting it; no matter, also, the speed the platform takes on from which that speed of light is being measured. The more such a platform approaches the speed of light, the more rulers used on it to measure distances contract and clocks used to measure time sequences go slower, so that on every occasion the self-same speed of light is registered. Later, in his General Theory of Relativity, Einstein further demonstrated that the accumulation of mass and energy generates gravitational fields through which space is being curved. In this scenario the planets follow a more or less straight path in the hollow space that is curved by the mass of the Sun.

It did not take long before the insight into the warped make-up of space-time was to result in the idea that the universe as such is subject to warping. Einstein's equations of the geometry of the universe predicted a universe that would either contract or expand. In order to obtain a universe in balance, Einstein introduced his cosmological constant. Independently of one another, however, both Alexander Friedmann in Russia and the Belgian priest and mathematician Georges Lemaître demonstrated that Einstein's original calculations gave evidence of an expanding universe. Einstein himself was so wedded to the idea of an eternal universe that he refused to accept this view until Edwin Hubble in 1929 proved that the universe was expanding. Hubble worked with one of the strongest telescopes of his time, which allowed him to look into galaxies beyond our Milky Way. When examining the spectra of stars in far distant galaxies he discovered that they had the same characteristic sets of missing colors as stars in our own galaxy, but that these were all shifted toward the red end of the spectrum. This meant that the galaxies were receding from us and from one another.

In 1931 Father Lemaître launched what is now called the "Big Bang theory." When running the expanding universe back in time, one had to eventually reach a point from which it all started: the "primeval atom," or "initial singularity" that gave rise to a terrific "fireball" from which not only the elementary particles came forth that later were to form the atoms and finally the clouds of gas and dust out of which the stars were born, but also the explosive expansion of space-time. Lemaître's picture of the Big Bang explosion was reminiscent of the first creation story in the Book of Genesis, according to which God separated light from darkness. Christian believers, among them Pope Pius XII, saw in it the affirmation of the power of the Creator who calls forth all things out of nothing.

Lemaître himself was more nuanced in his claims, realizing that his theory would create new problems. Einstein's Theory of Relativity perfectly predicts the behavior of large bodies in the macrocosm, but breaks down at the minuscule level of the "initial singularity." To describe the events close to this zero-point one must have recourse to quantum physics. Detailed studies of the quanta, however, show that they behave in a lawless manner. If one succeeds in measuring the position of a quantum (an electron, for example), then its velocity becomes uncertain and vice versa. The mysterious world of the quanta seems to be ruled by chance and contingency, so that only a probability calculus can tell us what is most probably going to happen. Particles may split into twin particles which, although travelling light years apart, may through instantaneous communication (faster than the speed of light) determine how to react to certain circumstances.

Moreover, particles sometimes behave as points and sometimes as waves. Their wave behavior may be such that interference occurs in which undulating waves either fortify one another or cancel each other out. In short, there is apparently no rule to predict the reactions of the quanta. When, from this background of indeterminacy, scientists set out to probe into the very "beginning" of the universe, they discovered that no clear-cut formula could be devised that determines the "mechanism" of the Big Bang.

That's why this book contains a lengthy chapter on quantum physics, one that may perhaps be challenging reading for those with a formation in the humanities. Yet, in order to acquaint oneself with recent theories concerning the origin of the universe—or of the multiverse, for that matter—a minimum knowledge of quantum physics is indispensable. One ought to get an idea of the specific characteristics of matter particles (fermions) and force-carrying particles (bosons), of antiparticles and quarks, of fields and strings, as well as of the typical effects of the four basic forces: the electromagnetic force, the strong nuclear force, the weak nuclear force, and the force of gravity. Only then can one begin to see the implications of the search for the merging of all these forces and elements at high energies, at the point close to the Big Bang in which once they were united.

In my account of the origin of the universe, I commence with the Hot Big Bang theory, launched in 1948. According to this theory the Big Bang, some 13.7 billion years ago, released a tremendous amount of heat. Whenever the universe doubled its volume, its temperature fell by half. This continuous drop in temperature gradually occasioned various symmetry breakings. These symmetry breakings had far-reaching consequences. At a certain moment this led to the preponderance of matter over antimatter (prior to it matter particles, such as electrons and neutrinos, and their related counterparts—positrons and anti-neutrinos—were caught up in a relentless process of mutual annihilation, with no possible chance for stable matter to emerge from it). Similar symmetry breakings caused the successive splitting off of the four forces. By means of these procedures, scientists were able to reconstruct the formation of atoms. Almost 100 seconds after the Big Bang atomic nuclei were formed (25% of them helium nuclei against 74% hydrogen nuclei). Yet, it would still take 380,000 years before these nuclei could capture electrons, so that complete helium and hydrogen atoms were born. These atoms coalesced into molecules, and these molecules would eventually, after millions of years, give rise to spiraling clouds of gas and dust. When, under the enormous impact of gravitational force, these spiraling clouds imploded, nuclear fusions took place in their inner core: hydrogen was burnt into helium, the procedure that makes the stars shine and produce energy.

More recent scenarios associate the emergence of the "initial singularity" with quantum fluctuations that, in one way or another, give rise not to a single universe, but to multiple universes. Similarly a more pivotal role is given to quarks—the newly discovered constituents of protons and neutrons—and their counterparts, anti-quarks, in the initial process of mutual annihilation. But more importantly, the Bang in the "Big Bang" is ascribed to the tremendous explosion, technically termed "inflation," that would have taken place a nanosecond after the eruption of the "initial singularity," and which caused our universe to almost instantly expand from one centimeter in diameter to ten million times the width of the Milky Way.

In all these scenarios, one cannot help but admire the genius of creative imagination: with the help of highly technical mathematical models and calculations, astrophysicists set out to reconstruct the various phases of the very beginning, in the hope that their theories may eventually be confirmed. For a good number of them the confirmations are spectacular. Between 1989 and 2010, ever more refined pictures were made available by NASA satellites of the Microwave Background that was released 380,000 years after the Big Bang (at the moment when the atoms were formed, so that space was vacated for photons to freely spread). In these pictures, temperature differences can be spotted that were the seeds that grew to become galaxies. And in March 2014 a research team at the South Pole reportedly discovered with their highly sophisticated telescopes specific patterns in the Microwave Background that are thought to be the effect of gravitational waves that must have been produced by the bursting forth of the exponential "inflation." Another spectacular event was the creation in July 2012 in the Large Hadron Collider (particle accelerator) in CERN, Switzerland, of the Higgs-field that bestows mass on the force-carrying particles and mass particles.

The above accounts of the formation of our universe are rather silent about the role of the Creator God in this rather stupendous account of our origins. This is also the case when they describe the cosmic processes that eventually led to the birth of life on earth. Biological life on earth could not have emerged if massive stars had not previously produced—through nuclear fusion and final catastrophic disintegration—carbon, oxygen, and iron, as essential ingredients of life. From this fact astrophysicists, like Stephen Hawking, conclude that our universe, and probably also a selection of other universes, are bio-friendly. But by no means does this imply that a Creator God would have been necessary to steer this process. To arrive at such a conclusion is, for them, excluded because of the many random processes of trial and error that are involved in the slow formation of the preparatory steps towards the appearance of intelligent life on earth.

It is at this juncture that the question arises of how to re-imagine the Creator God's relation to the emerging and developing universe. *Such a re-imagining is all the more imperative the more one realizes that the rock that shipwrecks all attempts at harmonizing an evolutionary universe with a Creator God is the classical theistic concept of a God designer, the God of a universe that is predetermined.* Yet, why stick to this concept of a prede-termining God, who rules out any chance factor in the make-up of the universe? *Times are ripe for theological imagination to conceive of a God who calls forth a universe endowed with an intrinsic creativity in which chance and random outcomes play an important role. The classic theistic God goes hand in hand with a static universe. Once, however, our universe is seen as continuously evolving, that static God must be replaced with a Creator God who is understood to be so generous that he allows the creation to share in his own creativity.*

Having completed this book I am grateful for the new insights I gained from it. At the same time, I would like to extend my gratitude to a number of persons who either made this project possible or were helpful in giving a finishing touch to its presentation. I thank Fr. Bienvenido Nebres, SJ, Dr. Antonette Angeles and Dr. Leovino Ma. Garcia for having invited me to teach this course at the Ateneo de Manila University. My gratitude also goes to the successive chairs of the Department of Philosophy, Dr. Rainier Ibana, Dr. Remmon Barbaza, and Dr. Agustin Rodriguez, who each in his own way encouraged me to carry on. Special thanks finally to Dr. Dennis Gonzalez who in 1992–1993 as a student in Leuven assisted me in the elaboration of the first sketch of this book, to Rev. Dr. David Gentry-Akin (Moraga, California, USA) for his generosity and competence in polishing the English of my text, to Fr. Herman Paulussen, SJ (Antwerp, Belgium) for verifying the accuracy of the section on theoretical physics, and to Dr. Edmund Guzman (Leuven, Belgium) for his technical assistance in producing the figures in the book. Thanks also to the many students whom I had the privilege of initiating in the amazing journey of cosmology.

CHAPTER I

The Creation Narrative in Genesis

The classic Jewish creation narrative, the so-called priestly version, figures in Genesis 1:1–2: 4a. Its opening verse is: "In the beginning God created heaven and earth." It was written after the Babylonian captivity, which ended in 539 BC. Its core message is that God, of his own choosing, created the world through the power of his commanding voice. The God of Israel and He alone is the sovereign creator, who calls forth things that previously did not exist, and whose whole existence continues to depend upon his power.

The Creating God Brings About Order

The calling forth of the creation through the sole power of God's word has in the course of history been encapsulated in the catch phrase "creation out of nothing" (*creatio ex nihilo*). Besides this standard view, however, a complementary reading can be defended that focuses on the Creator's act of imposing patterns of order upon a still chaotic world. In this reading the opening verse of Genesis, "In the beginning God created heaven and earth," is identified as the title of the book, whereas the creation narrative itself commences with the words: "*The earth was without form and void, with darkness over the face of the abyss, and a mighty wind that swept over the surface of the waters*" (Gen. 1:2).

In my comparison with Plato's *Timaeus* (see below), I will call this a "mythical secrecy." In order for the Creator to bring about order in the world and to assign to each entity its proper place, there must already, in a mysterious way, have been some scenario of disorder. The purpose of order is the dissolution of disorder. Even if one were to term this event in Plato's words as "order out of disorder" (*Tim.* 30a), this characterization would not militate against the "creation out of nothing"; for strictly speaking, "disorder" is not the material the biblical God uses to produce the world (this would result in a chaotic creation). What actually happened is that the Creator called forth the world "out of nothing" to be found in the mysteriously pre-given chaos. Through the power of his word the Creator generates a perfectly ordered world. He does so through a dramatic series of separations.

The first decisive separation God carried out is the separation of light from darkness ("let there be light"), followed by the separation of the waters above from the waters on the surface of the earth. What next comes is the separation of the waters on earth from the dry land, so that the stage is set for the various domains that will be populated with their proper entities. God attached lights to the vault of heaven: the sun for the day and the moon and stars for the night. God put clouds and birds in the air, and sea monsters and fish in seas and rivers. God made the land produce various species of plants and trees, and put animals (cattle, reptiles, wild animals) on it. Finally, God crowned all these works with the creation of the human species, male and female, made in the divine likeness.

This listing may draw one's attention to the created things, yet the real focus is announced by the solemn opening sentence: "In the beginning God created heaven and earth [and all that these realms contain]." God alone is the sovereign Creator. This pivotal insight comes to the fore in the various stages through which God decides to create: "*God said*: let there be light," "*and God said*: let there be a vault between the waters," "*and God said*: let there be lights in the vault," "*and God said*: let the waters teem with countless living creatures." In short, God's creative word calls into being all things that are essential to the flourishing of life on earth. This gesture is accompanied by the refrain: "*And God saw that it was good.*" Good are the rhythms of day and night, and of rain and dryness. God's blessing rests on the food chains (animals and human beings are allowed to eat plants; human beings rule over birds, fish, and animals) and on the procreation of all the species. Once this multifarious continuity has been ascertained, God can afford to "rest" in the glory of creation, which reproduces itself under the sign of divine affirmation and blessing.

As regards the crowning element, the humans, their multiplication from generation to generation, "is good." God's special blessing rests upon them: God created them in the Creator's own image, and created them male and female. Moreover, the second creation narrative (Gen. 2: 4–25), which in fact is much older than the priestly version, tells us that the Lord God "formed the body of the human (Adam) from the dust of the ground and breathed into his nostrils the breath of life." Eve partook of this life-giving breath when she was formed from Adam's rib. The human species is called to partake with the sovereign God in the mastery over creation. Humanity is allowed "to rule over the fish in the sea; the birds of heaven, and every living thing that moves upon the earth." Unlike the animals, however, the human beings are going to engage in agriculture.

Genesis and Exodus

In the early history of humankind, one can discern various stages. Initially, the humans were food collectors and hunters. Only 10,000 years ago did they begin to settle down in villages, cultivate fields, and raise cattle. Then around 3000 BC, there arose urban centers and cultural monuments in Mesopotamia, Egypt, and various parts of Asia. The oldest parts of the book of Genesis most probably began to take shape after the Israelites' exodus from Egypt, which some scholars tentatively place between 1290 and 1224 BC. So, the Israelites must have been acquainted with urban cultures: in Egypt and much later also in Babylon. Yet, the two creation narratives in Genesis clearly suggest a rural setting. For example, "God said: I give you all plants that bear seed everywhere on earth, and every tree bearing fruit which yields seed; they shall be yours for food. All green plants I give for food to the wild animals, and to all the birds of heaven, and to all reptiles on earth, every living creature" (Gen. 1:29).

In the second creation narrative, which also contains "the fall and expulsion from Paradise," the rural context is even more conspicuous. The narrative begins with the observation that "when the Lord God made earth and heaven, there was neither shrub nor plant growing wild upon the earth, because the Lord God had sent no rain on the earth; nor was there any human to till the ground" (Gen. 2:5). Furthermore, it is said that the Garden of Eden is an agricultural paradise encircled by fecund rivers. Once expelled from this paradise, however, human beings will have to labor and to earn their food from a land full of thorns and thistles (3:17–19). Astronomical data that were already available from the surrounding cultures are not given much attention. The lights in the vault of heaven are only there to mark the festivals and the seasons; they are not subjects of further curiosity.

Given this agricultural setting, one would have expected a more pronounced veneration of the natural forces (e.g., springs, mountains, and rivers), but this is apparently not the case. For the Jews, nature cannot be an object of worship, because God is the Lord of history especially as regards the astonishing account of the making of the Israelite nation. God is primarily the God of the Exodus, the One who calls forth a new future for

Israel out of nothing within the existing state of affairs. This event became the prototype on which the Jewish creation narrative was modeled. Thus, all that Genesis has to say about the blessing (or the curse) of the agrarian milieu was put within the framework of the God who makes all things new by triumphantly banning the existing chaos. Just as the Hebrews were kept in captivity in Egypt, so too were the cosmic powers kept in captivity, so to speak, until they were set free to adopt a new decisive order through the power of God's word. In the act of creation, God called forth the world's true future out of nothing [see 2 Macc. 7:28] just as the Creator has shown divine power in the liberating act of the Exodus event. God's Word created things that were not there in an ordered way: the starry skies, the dry land, the seas, the fish in the waters, the plant and animal species on earth, and finally the human (specifically the Jewish) ancestors—all of them in due time.

Credit, therefore, is due to those authors who, in terms of literary dependence, draw a comparison between the opening verses of *Genesis* (the taming of the chaotic elements) and the liberating march through the Red Sea (where God also tames the waters).[1] Let us recall again the salient texts of the priestly version: "In the beginning God created heaven and earth. *The earth was without form and void, with darkness over the face of the abyss, and a mighty wind that swept over the surface of the waters.*" And then it happens, "God said: 'let there be light,' and there was light . . . *God separated light from darkness*" (Gen. 1:1–4). Much later in the text, "God said: 'let there be a vault between the waters, *to separate water from water.*' So God made the vault, and *separated the water under the vault from the water above it*" (vv. 6–7). "And God said: 'let the waters under heaven be gathered into one place, *so that dry land may appear,*' and so it was" (v. 9). God not only tames the chaos but also does so in such a way (the powerful separations!) that it reminds us of the great event of the march through the Red Sea, where JHWH split the waters. According to Exodus 14:21: " Moses stretched out his hand over the sea, and the Lord *drove the sea away* all night *with a strong east wind* and turned the sea-bed into *dry land.*" The dry land that arose from the tempestuous waters gave witness to God's creative power both in the order of nature (Genesis) and in the order of the history of salvation (Exodus). Just as the emergence of a path through the sea inaugurates freedom and a promising future for Israel, so too the

---

1   See Gerhard von Rad, *Old Testament Theology,* Vol. I, trans. D. M. Stalker (New York: Harper and Brothers, 1962), 139.

splitting of the waters by virtue of God's potent "word in the beginning" discloses the path for vegetable and animal life to grow.

Isaiah 51:9–10, written during the Babylonian captivity, is a fine example of the way in which Jewish thinking likes to merge the imagery of creation and Exodus:

> Awake, awake, put on your strength, O arm of the Lord, awake as you did long ago, in days gone by. Was it not you, who hacked the Rahab [the mythical sea monster] in pieces and ran the dragon through? Was it not you who dried up the sea, the waters of the great abyss, and made the ocean depths a path for the ransomed?

This text evokes, in the midst of historical setbacks, the belief in God the Creator who liberates. Rahab is the chaotic sea monster that represents the power of Egypt. The sea, which according to the creation narrative is being separated from the dry land, is associated with the Red Sea that dries up so that Israel can escape from the house of its captivity.

God's Transcendence vis-à-vis the World

If JHWH was not transcendent to history and the created realm, He would not have been able to steer the Israelites in accordance with the divine will. JHWH is the sovereign Lord of creation whose power is never toppled by the vicissitudes of history. The Lord God always remains the inexhaustible source of creativity whose distinct sovereign existence exceeds the being and activity of the created entities; these only exist through God's compelling word without which they would collapse into nothingness. The created realm only reproduces itself—as it ought to be—in virtue of God's commanding voice.

No antecedent causes can condition God's free acting, for the divinity alone lays down the rules of Its creative power and of the manner in which it should be worshipped. Whoever is fascinated by the beauty of creation to the point of forgetting thereby to praise the glory of the Creator would by the same token abruptly drift away from the good order of creation. Such an idolater is going to be chased pitilessly from the Garden of Eden. This is the penalty for those who place created things and the Creator on the same footing. In this light, one ought to understand the polemics waged against the idolatry of natural forces which some authors discover in the Jewish creation narratives. The German theologian Jürgen Moltmann comments on this as follows:

The first sentence of the Priestly Writer's creation narrative is the summing up of a long process of reflection by Israel's faith. Because this thinking belonged within the framework of the dispute between faith in Yahweh and the cosmogonies of the religious cults from Egypt to Babylon, this opening sentence reflects a deliberate confrontation. The world, we are told here, is not the result of a struggle between the gods, as the Enuma Elish epic says. Nor was it born from a cosmic egg, or from some primordial matter. To say that God "created" the world indicates God's self-distinction from that world, and emphasizes that God desired it. This means that the world is not in itself divine; nor is it an emanation of God's eternal being. It is the specific outcome of his decision of will. Since they are the result of God's creative activity, heaven and earth are neither divine nor demonic, neither eternal like God himself, nor meaningless and futile. They are contingent. They are his goodly work in which he has pleasure—no more than that, but no less than that either. They take their reality from the affirmation by their creator.[2]

Moltmann further remarks that the Hebrew verb *bara'* (to create) is used exclusively as a term for the divine bringing forth through which something comes into existence that was not there previously:

Bara' is never used with the accusative of a material out of which something is to be made. This shows that the divine creativity has no conditions or premises. Creation is something absolutely new. It is neither actually nor potentially inherent or present in anything else. The text makes a clear distinction between "creating" (*bara'*) and "making" (*'asah*). In Gen.1:1, the word "create" is used for creation as a whole. The "making" begins in v.2, as it were, and is completed with the Sabbath: "So on the seventh day God finished his work which he had made." (Gen. 2:2)[3]

The verb *'asah* is not typical of God alone; human craftsmen may also "make" something precious, but applied to God, *'asah* expresses something more solemn. It is striking, however, that *'asah* is being used for the "making" of the heavenly bodies, whereas *bara'* is used for conveying the creation of human beings.

Reflecting on these points, Moltmann concludes that

the later theological interpretation of creation as *creatio ex nihilo* is therefore unquestionably an apt paraphrase of what the Bible means by "creation." Wherever

---

2　　Jürgen Moltmann, *God in Creation: An Ecological Doctrine of Creation* (London: SCM-Press, 1985), 32.

3　　Ibid., 33.

and whatever God creates is without preconditions. There is no external necessity which occasions His creativity, and no inner compulsion which could determine it. Nor is there any primordial matter whose potentiality is pre-given to his creative activity, and which would set him material limits.[4]

Let us recall that the "creation out of nothing" was explicitly defined a dogma of the Catholic Church in the Fourth Lateran Council (1215), and in Vatican I (1870).

The Question of Secondary Causes

The theologian's main task is without doubt to transmit the God-centered message of the biblical creation narratives, yet this is only one part of the enterprise. Theologians and believers today are also expected to integrate recent findings concerning the origin of the universe and life on earth. That to which Genesis invites us, viz., the admiration of God's good creation with its beautiful rhythms of reproduction, is going to lack depth if it is not sustained by a minimal knowledge of how nature actually works. To emphasize only God's absolute self-distinction from the world, as Moltmann does in line with the Protestant tradition, threatens to close the believer's eyes to the wondrous working of that very world. It fails to do justice to a "natural religiosity," the rediscovery of which is so much needed today. This realization, e.g., of the amazing precision of constants of nature or of natural laws, can make us sensitive again to the splendor of nature and to the grandeur of nature's Creator.

Prior to the Reformation, Christian theologies frequently concerned themselves with the wondrous workings of nature in their considerations about the secondary causes that, with the support of the primary cause (God), are operational in maintaining the good order of creation. In his *Commentary on the Book of Causes*, a Neo-platonic compendium written by Proclus, Thomas Aquinas—in line with Greek cosmology (see below)—acknowledges the import of secondary causes in creation. These are, so to speak the instruments the Creator uses to achieve a precise purpose: "God, the primary cause, is the unmoved mover, whereas the things he sets in motion—the celestial spheres—are the instruments, the secondary causes, he employs to achieve his will and execute his plans."[5] The rotations of the celestial spheres have a bearing on the life cycles on earth.

---

4   Ibid., 34.

5   Max Wildiers, *The Theologian and His Universe: Theology and Cosmology from the Middle Ages to the Present* (New York: The Seabury Press, 1982), 56.

It is obvious that Christian reflection on the secondary causes could not have developed without a steady assimilation of earlier Greek cosmologies into the Christian faith. Indeed, it is a firm tradition in patristic and medieval theology to harmonize Genesis with the (then accepted) scientific theories about the order of nature, viz., Plato's *Timaeus* and Aristotle's *Metaphysics* and *De Coelo*. As a result of his protestant background, with its characteristic emphasis on "scripture alone," Jürgen Moltmann has dropped any reference to this tradition, which nonetheless offers a platform from which to engage in a dialogue with later scientific explanations of the workings of the universe. Why should not Newton's law of gravity or the subatomic "fine tunings" be regarded as secondary causes in creation?

Furthermore, it should be noted that the Protestant canon of the Hebrew Bible omits important passages of Wisdom literature that seem to have prepared the notion of secondary causes in creation. One ought only to think of the way in which the Book of Jesus Sirach describes the role of the rainbow and the host of stars:

> The glory of the stars beautifies heavens, the stars being the luminous ornaments in the heights of the Lord. Upon the Word of the most Holy they carry out their task, staying alert on their station. Look at the rainbow and bless its maker; it is so glorious and beautiful, forming a heavenly arch of splendor, bent by the hands of the Almighty. (Sir. 43:9–10)

CHAPTER 2

Plato's Creation Narrative in the Dialogue *Timaeus*

In the Greek world, Plato's creation narrative in the dialogue *Timaeus* provides a complement to the biblical narrative; it differs from the latter both in style and in content. The Priestly version of *Genesis* is condensed, rhythmic, and liturgical rather than philosophical, and is addressed to the community of the believers. Plato's narrative, on the other hand, continues a rich Greek tradition that combines philosophy of nature and cosmogony, and addresses itself to a learned audience.

Cosmology and Theology

The *Timaeus* is far from being a secular work. One ought to realize that, in the Greek milieu, both the philosophy of nature and the reflections on the genesis of the cosmos were part of the discourse on the gods (theology). The Greeks had their own narratives or theogonies, which contained genealogies of gods and heroes (demigods), yet to understand the meaning of God and the Divine, they had recourse to the study of nature (*physis*: the things that grow and develop). In their eyes, to say something meaningful about God, one has to start by investigating the fundamental structures of whatever exists and has its origin in God.

It is no surprise, thus, that Plato in his mature period, after having treated topics of ethics, epistemology, and politics, set out to tackle cosmological problems. In doing so, he borrowed insights from various authors before him such as the Ionic philosophers of nature (6th century BC), Empedocles (483–423), Anaxagoras (499–428), and above all Pythagoras (580–500). Owing to Thales of Miletus (610–547), Plato (428–347) adopted the view that all things can be resolved into some most simple element, whereas Empedocles had taught him that the material world is made up of four elements (fire, air, water, and earth), and Anaxagoras that the

World-Artificer, after having imposed general order on a pre-given chaotic matter, charged the stars and the heroes to complete with the creator the details of this enterprise. On the whole however, Plato drew his inspiration mainly from Pythagoras, who had already ventured to explain the structures of the cosmos in terms of mathematical and musical proportions.

In the first chapters of the *Timaeus*, Plato uses the style of a dialogue. Looking at these chapters, we find, in addition to Socrates, three guest speakers invited to a symposium. One of them, Timaeus of Locri in southern Italy, is the first to take the floor. Southern Italy, a Greek colony, was at that moment the important centre of Pythagorean cosmological studies.

> Consider now, Socrates, the order of the feast as we have arranged it. Seeing that Timaeus is our best astronomer and has made it his special task to learn about the nature of the universe, it seemed good to us that he should speak first, beginning with the origin of the Cosmos and ending with the generation of mankind. After him I am to follow.[1]

The religious aura of cosmology is evident from the way Socrates invites the speakers to invoke the gods before the discussion is going to start. Socrates presents this invocation as something quite natural to which Timaeus responds:

> Nay, as to that, Socrates, all men who possess even a small share of good sense call upon God always at the outset of every undertaking, be it small or great; we therefore who are purposing to deliver a discourse concerning the universe, how it was created or haply is uncreate, must needs invoke Gods and Goddesses (if so be that we are not utterly demented), praying that all we say may be approved by them in the first place, and secondly by ourselves. (*Tim.* 27cd)

The Good Creator and the Beauty of the World

The *Timaeus* is one of Plato's mature dialogues; it presupposes some basic tenets of the earlier dialogues, such as the existence of eternal ideas and the immortality of the soul. Platonism methodically separates the realm of archetypal forms from their inadequate realization in the material world. True knowledge (*episteme*) derived from the eternal ideas prevails over opinions gathered from sense impressions (*doxa*). Empirical knowledge is not entirely reliable because it comes forth from changeable things,

---

1   Plato, *Timaeus*, trans. R.G. Bury, The Loeb Classical Library, IX (London: Heinemann, 1929 [1966]), 27a.

whereas true knowledge results from the study of stable forms and harmonious proportions.

This basic principle has a bearing on the cosmological enterprise. Since the observable world is subject to change, it cannot yield any insight into the basic structures of the universe. These structures must be meta-empirical in nature, and are best rendered by mathematical and geometrical proportions. On the basis of these proportions the cosmos will be beautiful, and Plato goes on, this shows that its Maker is good and without envy. For his desire was that the things that were going to come into existence would be "so far as possible, like unto Himself" (*Tim.*, 29e).

In his creation narrative Plato stages three major actors: the divine Demiurge, the eternal ideas, and the already-existing matter, which is the receptacle (formative of space) of the eternal forms. These three actors are present in the very beginning. Added later are two further actors: the World Soul and the planets.

The story commences with the Demiurge fixing his gaze on what is self-identical, on the eternal ideas after which the material world is going to be structured. Plato finds this gesture of gazing at the perfect model so important, that he concludes from it that the Maker must have acted out of sheer goodness. He writes:

> Now to discover the Maker (*ton poieten*) and Father (*ton patera*) of this Universe were a task indeed; and having discovered Him, to declare Him unto all men, were a thing impossible. However, let us return and inquire further concerning the Cosmos,— after which of the Models did its Architect construct it? Was it after that which is self-identical and uniform, or after that which has come into existence? Now if so be that this Cosmos is beautiful and its Constructor (*demiourgos*) good, it is plain that he fixed his gaze on the Eternal; but if otherwise (which is an impious supposition), his gaze was on that which has come into existence. But it is clear to everyone that his gaze was on the Eternal; for the Cosmos is the fairest of all that has come into existence and He the best of all Causes. (*Tim.* 28c–29a)

It is here that the "third actor" will come into play: the receptacle—chaotic matter—that is already present before the Demiurge begins to bring it into order: Seeing that matter "was not in a state of rest but in a state of discordant and disorderly motion, He brought it into order out of disorder, deeming the former state is in all ways better than the latter" (*Tim.* 30a). The procedure the demiurge used to bring matter into a state of order is quite simple: he sunk the imprint of the eternal ideas into matter, so that it became "ordered matter."

This allows us to perceive some elements of similarity and difference between *Genesis* and Plato. (i) Just as in *Genesis,* the God-creator (Demiurge) is the one who imposes order on things, though the manner of doing so is different. In the bible, God brings about order through the power of his word, whereas in the *Timaeus,* the Demiurge does so by previously fixing his gaze on what is self-identical, on the eternal ideas after which the world will be modeled. Plato's demiurge can thus rightly be termed an architect who through his knowledge of mathematics and geometry lays down the foundations of the world as a structured whole. (ii) In *Genesis,* too, reference is made to an "earth without form and void" and to a "darkness over the face of the abyss" before God lets shine light on it. One may call this a *mythical secrecy,* in as far as chaos is needed to produce order. Plato's *Timaeus* is no exception to the rule. Matter, the initially disordered element, is already there. But, unlike in *Genesis,* this disordered element begins to strive for receiving the imprint of the archetypal forms. The material entities in the world would not yearn to be shaped by the imprint, if there were no geometrical forms or sets of proportional ratios capable of giving a structure to matter. Still, there is more to it. Taken in itself matter would not be striving for order. To promote that striving, a vivifying World Soul must be in place.

The World as a Living Organism: The Role of the World Soul

The World Soul's task is to arouse the body of the universe to crave for the penetration of the intelligible forms and their proportionate combinations. Without this animating force, the world will lack a true receptivity for eternal beauty. It is no surprise then that the *Timaeus* sets great store by the World Soul as *the* immanent cosmic principle. The World Soul ought to be constructed with special care, for only that which is entirely proportional in form is able to lure the world towards splendid patterns of order. For clarity's sake, I will render Plato's account of the construction of the World Soul in a later section, and instead explain first the working of the World Soul on the material elements of the world, that is, on the body of the world.

Plato's creation narrative "vacillates continually on the border-line between mathematics and physics, between the realm of pure thought and that of natural things between myth and empirical reality, between fancy and facts."[2] We have seen this mythical element at work in the way the Demiurge fixes his gaze on the mathematical structures he would like to impose on disordered matter. A similar mythical background transpires in

---

2   E.J. Dijksterhuis, *The Mechanization of the World Picture*, trans. C. Dikshoorn (New York: Oxford University Press, 1969), 17.

his presentation of the world as "a living organism endowed with soul and mind" (*zoon empsychon ennoun*: Tim. 30b). The mythical genius likes to work with images and metaphors, which are plausible on a *prima facie* consideration but which do not stand the test of reason. Properties within such disparate domains as nature and the human culture are being perceived as if they had a common, homologous structure. Thus, a person's life span is likened to the rhythm of the seasons. The transference of properties from one domain to another creates a common bond between them. Mythical language elaborates a "gigantic hall of mirrors in which the humans and the world reflect homologous aspects back into each other, in spite of contrasts."[3] The humans apparently need these homologies to find themselves at home in nature. In this light one ought to assess Plato's mythical talk about the world as a living organism. We ourselves, as well as all inanimate things, belong to this organism.

For Plato, this mythical language serves as a springboard to jump to scientific explanation. As we will see, this explanation focuses on the means proportional that are needed to harmonize geometrical configurations. In line with Empedocles, he tells us that the body of the world is made up of fire, air, water, and earth. For a material thing, he says, to be visible and tangible it must contain fire and earth; in order to bind these two elements together, two means proportional are required because, unlike a two-dimensional surface, a solid possesses depth or cubic extension also. That is the reason for the interposition of two other elements (air and water) between fire and earth in accordance with a particular arithmetical ratio. Plato writes: "Thus it was that in the midst between fire and earth God set air and water, and bestowed upon them, so far as possible, a like ratio one towards another—air being to water as fire to air, and water being to earth as air to water" (*Tim.* 32bc).

It is not by enumerating the four elements that one already knows how they are solidly kept together. To get an insight into the cementing bond that glues the four elements together, one has to have recourse to a system of proportionalities like that elaborated by Pythagoras.

The system Plato is alluding to is the proportional arithmetical series of the type a:b = b:c = c:d (**a** is to **b** as **b** is to **c**, and as **c** is to **d**). Take, for example, in a simplified form, 2:4 = 4:8 (two is to four as four is to eight); the pivotal element here is the middle term 4 between the two extremes 2 and 8. This middle term is called in Greek: *mesotes* or *mean proportional*. What this *mean proportional* is about can be illustrated as follows: imagine you have

---

3   Maurice Godelier, "Mythos und Geschichte," in *Die Entstehung von Klassengesellschaften*, ed. Klaus Eder (Frankfurt: Suhrkamp, 1973), 316 (translation mine).

two posters, one of 20 cm x 20 cm (400 cm^2), and another of 80 cm x 80 cm (6400 cm^2). You set out to hang them on a wall and to keep them apart by a distance of three meters. This gives a rather odd result in terms of spatial proportion. In order to eliminate the disharmonious effect, you will have to add another poster between them, a third poster whose width and length is the middle term of the widths and lengths of the two extreme posters. Only an intermediate poster of 40 cm x 40 cm (1600 cm^2) is able to bridge the gap between the disproportional sizes of the small and big posters. The whole is now fully harmonious, for 400 cm^2 is to 1600 cm^2 as 1600 cm^2 is to 6400 cm^2. [You can even reverse the order and say that 6400 cm^2 is to 1600 cm^2 as 1600 cm^2 is to 400 cm^2, and even that 1600 cm^2 is to 400 cm^2 as 6400 cm^2 is to 1600 cm^2.] Indeed, the proportions are from various sides harmoniously attuned to one another.

Plato's concern, now, is about bridging the spatial distance between two extreme *solids*, which are three-dimensional, and not just between two extreme posters with a flat surface. In the case of three-dimensional objects, we have the additional dimension "depth," which the flat posters fail to have. Therefore, *two* means proportional are needed to bridge the gap between the two extremes. Take an Egyptian pyramid and the Kaaba of Mecca, which for simplicity's sake we can consider a perfect cube. To bring about a gliding transition between those two solids, one needs two interposed solids whose forms obey the principle of the mean proportional. Let the pyramid be **a** and the Kaaba be **d**; the proportional series representing the transition would then be the following: a:b = b:c = c:d, whereby **b** and **c** are provisionally two unknown magnitudes. To identify them, one has to consult the findings of geometry—the branch of mathematics concerned with the properties and relations of points, lines, surfaces, solids, and higher dimensional analogs— and stereometry, which is concerned specifically with the measurement of solid bodies.

The Kaaba of Mecca (cube) is made up of six square planes, whereas the pyramid is made up of four equilateral triangles (three of which meet at the top, and one of which forms the basis). Now, geometry tells us that a square plane can be divided into two rectangular isosceles triangles, each of which can in turn be divided further into two triangles of the same kind. What is first to be done, thus, is to decompose the plane faces of the cube and the pyramid into their more simple constituents: smaller rectangular isosceles triangles, rectangular scalenes (through which other figures can be made). After this decomposition, one has derived the materials from which to build the missing intermediate solids b and c. Handbooks of stereometry contain the triangle combinations that can produce prisms, octahedrons, dodecahedrons, etc., and also which solids can be partially or approximately converted into others (like pyramids into cones).

Plato examines in detail how the plane faces that are used to construct the geometrical solids of the four elements can be resolved into various triangles:

> In the first place, then, it is plain, I presume, to everyone that fire and earth, and water and air are solid bodies; and the form of a body, in every case, possesses depth also. Further, it is absolutely necessary that depth should be bounded by a plane surface; and the rectilinear plane is composed of triangles. Now all triangles derive their origin from two triangles, each having one angle right and the others acute. (*Tim.* 53cd)

The two triangles are (a) the rectangular isosceles, and (b) the rectangular scalene, with the help of which equilateral triangles can be formed. What matters, then, is to see

> what will be the four fairest bodies, dissimilar to one another, but capable in part of being produced out of one another by means of resolution [into constituent parts]; for if we succeed herein we shall grasp the truth concerning the generation of earth and fire and the means proportional. (*Tim.* 53e)

After this, Plato explains what geometrical solid is paired with which particular element. Fire, which moves upward, must have the form of a rising *pyramid*, a solid whose base is, in this case, a square and whose other faces are equilateral triangles with a common vertex. Fire in Greek is **PYR**; a burning fire takes on the form of a pyramid. The element earth, however, must be something heavy and relatively stable; therefore, it is best represented by a *cube*, a solid bounded by six plane faces of equal area and making right angles with one another (the six plane faces can be resolved into triangles, though not of the kind of equilateral ones).

Air takes on the form of an *octahedron*, a solid with eight plane faces each of which are equilateral triangles arranged in such a way that the resulting form is two pyramids joined together at their base (one pyramid put on its head, and the other with its top in the air). Water, finally, takes on the massive form of an *icosahedron*, a solid with twenty sides. This figure can be likened to the Pentagon in Washington, DC, but whereas the Pentagon has five rectangular walls, the icosahedron's "walls" are made up of ten equilateral triangles whose tops are alternately up and down. This structure is then bounded above and below by two rather low pyramids (each of them

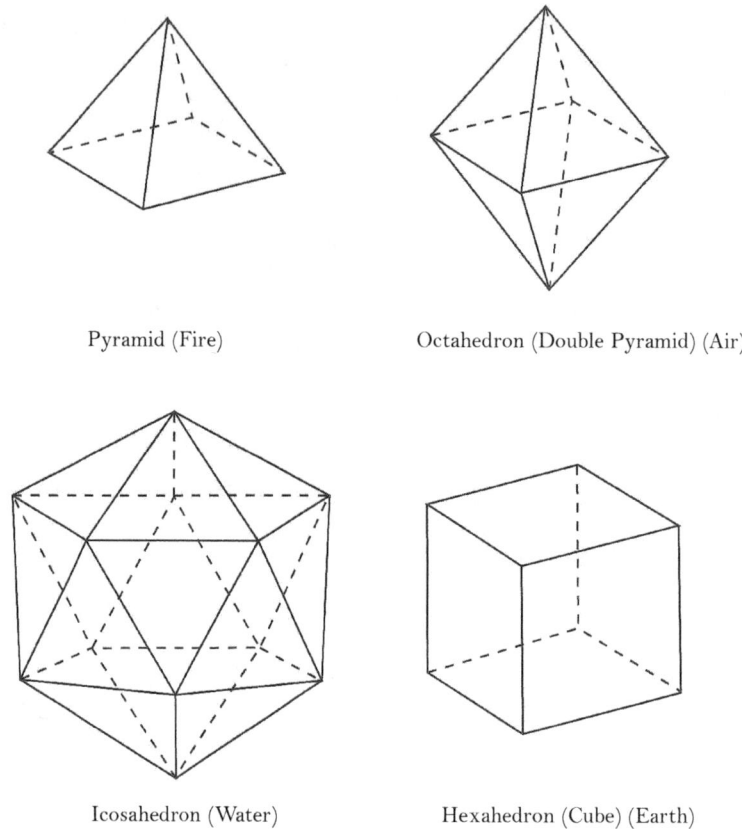

Pyramid (Fire)   Octahedron (Double Pyramid) (Air)

Icosahedron (Water)   Hexahedron (Cube) (Earth)

Figure 1. Four Platonic solids

having five equilateral triangles with a common vertex); one pyramid is on
top of the pentagonal structure, and the other is inversely suspended from
the bottom of the structure. This way, the icosahedron presents itself as a
transition between the double pyramid, "air," and the cube, "earth."

Pyramid, double pyramid, pentagonal structure plus double pyramid,
and cube[4] manifestly are forming a series in which two intermediate solids
constitute the transitional links (*mesotetes*) between two extremes. Indeed,
rising fire (pyramid) is bonded to the heavy cube "earth" by means of the

---

4   Besides the four regular solids representing the four elements, Plato mentions a fifth
     regular solid, the dodecahedron, made up of 12 pentagons (five-side faces) and 20 vertices
     (arranged in such a way that three pentagons meet at each vertex). This dodecahedron
     represents the 12 signs of the Zodiac, and is sometimes associated with quintessence (see
     further: analysis of Aristotle).

double pyramid "air" and the water drop of the pentagonal structure with its additional pyramids above and below. So in the three-dimensional space, the four elements are, as far as possible, bound together through a progression of proportional solids: a:b = b:c = c:d. A more perfect bond is hardly possible; it reveals the wonders of nature's architectonic harmony.

Not only are the elements proportionally joined together, they also admit the possibility of being transformed from one to another. Observation shows that water can be converted into steam (air) and that air is under certain circumstances flammable. Moreover, Plato invites his readers to imagine these elements as minuscule particles; this shows that he would have been prepared to grasp the contemporary insight into the means by which chemical elements can be transformed into one another.[5] He opines that one unit of water (icosahedron) is convertible into one unit of fire (pyramid) and two units of air (double pyramid), and that one unit of air can be resolved into two units of fire, whereas 2.5 units of air can be condensed into one unit of water. Plato remains consistent with himself when reporting that here the traceable series of transformations stops: indeed, the element earth (cube) stands somewhat apart in that its ultimate constituents—rectangular isosceles triangles—are apparently not convertible into equilateral triangles. This "unevenness" forebodes the *leimma* (rest) with which he grapples in the reconstruction of the World Soul.

The Musical Construction of the World Soul

The explanation of how the material elements are bonded to one another by harmonious proportions is in line with Plato's account of the composition of the World Soul. Right from the outset, Plato speculates on the global form of the cosmos. The world must be self-sufficient, i.e., must have no hands or feet for receiving some help from outside. So, its form must be spherical: a sphere moving itself in virtue of an inner principle, the World Soul.

In fact, the Demiurge constructs the World Soul prior to establishing order within the material elements of the world; only for clarity's sake have I treated the harmonious conjunction of the four elements first, though Plato deals with it later in the *Timaeus*. This priority is logical: the more perfect ought to precede the less perfect. The World Soul excels all other things in dignity. She is the vivifying force animating the world from

---

5   See Alfred E. Taylor, *A Commentary on Plato's Timaeus* (London: Clarendon Press, 1972), 382–83.

within, stretching from the interior of the earth to heaven, where she is most visibly shown:

> The ever existing God made the body of the world smooth and even and equal on all sides from the centre, a whole and perfect body compounded of perfect bodies. And in the midst thereof He set Soul, which He stretched throughout the whole of it, and therewith He enveloped also the exterior of its body; and as a Circle revolving in a circle He established one sole and solitary Heaven, able of itself because of its excellence to company with itself and needing none other beside [...]. And because of all this He generated it to be a blessed God. (*Tim.* 34b)

With particular care the Demiurge sees to it that the World Soul is composed of harmonious parts. He hereby acts like a real craftsman, kneading and fashioning the World Soul's proportions just as the formation of Adam's body by the biblical God. The Demiurge's specific concern is to unite within one whole the characteristics of the indivisible (the self identical, the same) and the divisible (the other), for the World Soul ought to be able to communicate unity (ideal forms) to the material realm of multiplicity. The reader is invited to pay attention to Plato's meticulous description of this blending:

> Midway between the Being which is indivisible and remains always the same and the Being which is transient and divisible in bodies, He blended a third form of Being compounded out of the twain, that is to say, out of the Same and the Other [...]. And He took the three of them, and blent them all together into one form, by forcing the Other into union with the Same, in spite of its being naturally difficult to mix. (*Tim.* 35ab)

This process of blending is only the prelude to the Demiurge's careful repartition of portions taken from the mixed whole: each portion corresponding to a specific number or mathematical ratio. The mythical talk about the Demiurge acting in person is, again, the springboard to jump to mathematical (and in this case) also musical considerations. For the portions the Demiurge is to lay out, the one next to the other, will eventually turn out to be the markers of the whole tones and semi-tones formative of tetrachords and octaves.

Plato goes on:

> And when with the aid of [the third kind of] Being he had mixed them, and had made of them one out of three, straightaway He began to distribute the whole thereof into

so many portions as was meet; and each portion was a mixture of the Same, of the Other, and of [the third kind of] Being. And he began making the division thus: First he took one portion from the whole (1); then He took a portion double of this (2); then a third portion, half as much again as the second portion, that is: three times as much as the first (3); the fourth portion he took was twice as much as the second (4); the fifth three times as much as the third (9); the sixth eight times as much as the first (8), the seventh twenty seven times as much as the first (27). After that, He went on to fill up the intervals in the series of the powers of 2 and the intervals in the series of powers of 3 in the following manner. He cut off yet further portions from the original mixture, and set them between the portions above rehearsed, so as to place two Means in each interval—one a harmonic Mean [...]; the other an arithmetical Mean [...]. And whereas the insertion of these links formed fresh intervals in the former intervals, that is to say, intervals of 3:2 and 4:3 and 9:8, He went on to fill up the 4:3 intervals with 9:8 intervals. This still left over in each case a fraction, which is represented by the terms of the numerical ratio 256:243. And thus the mixture, from which He had been cutting these portions off, was all spent. (*Tim.* 35b–36b)

Let us first focus on the rough division into parts that can be quantified as 1, 2, 3, 4, 9, 8, 27. Within these series of whole numbers, one may discern two types of progressions: a progression of the powers of two (2, 4, 8) and a progression of the powers of three (3, 9, 27). This leads to the combined formula: 1, **2**, 3, **4**, 9, **8**, 27 (the series of the powers of two is in bold type). Again we have to do with proportions (2:4 = 4:8) and with a mean proportional (4). The same is true for the series of the powers of three: 3:9 = 9:27, whereby 9 is the mean proportional. Instead, however, of using this basic framework for further mathematical considerations, Plato will use it, in line with Pythagoras, for constructing a musical melody: the music of the spheres.

Greek music is based on the tetrachord, a group of four notes or half an octave. Unlike other European music, this group of notes takes on a descending melody: *la*, sol, fa, *mi* (Doric tetrachord). By putting another tetrachord before this one, one gets a full Doric scale: *mi*, re, do, *ti*, *la*, sol, fa, *mi*. Besides the Doric mode, there were also others, like the Phrygian mode: *re*, do, ti, *la*, sol, fa, mi, *re* or the Lydian mode *do*, ti, la, *sol*, fa, mi, re, *do*. The melodies based on these descending scales clearly differ from ours. One ought to think only of Theodorakis' *Canto General*, and to compare it to Beethoven's *Hymn of Joy*. Theodorakis' descending melodies give an epic character to his composition, whereas Beethoven expresses his joy through an ascending melody. Let us note that a violin has four strings; upon each of them a tetrachord can be played.

What Plato aims at is the construction of various pairs of tetrachords (or their respective octaves) on strings that are already marked by the measurements 1, **2**, 3, **4**, 9, **8**, 27. To do so, he interposes between the mentioned integers two Means proportional, one harmonic (in bold)—based on the multiples of 1/3—and the other arithmetical—based on the multiples of 1/2. So, the interval between 1 and 2 is filled up as follows: 1, $1^{1/3}$, $1^{1/2}$, 2; that between 2 and 4 as 2, $2^{2/3}$, 3, 4. Plato's contemporaries would already have understood that with the insertion of a harmonic and an arithmetical Mean, he had formed a whole tone, which in Greek music is longer than a western whole tone because it has a ratio of 9/8. Indeed $1^{1/2}$ or 3/2 divided by $1^{1/3}$ or 4/3 yields the ratio 9/8.

In a next step, then, Plato sets out to fill up the newly created intervals: that between a basic integer and the inserted harmonic Mean (between 1 and 1⅓ or 4/3; between 2 and $2^{2/3}$ or 8/3 etc.), and that between the inserted arithmetical Mean and the subsequent basic integer (between $1_{1/2}$ or 3/2 and 2, and between 3 and 4, etc.). If we term these new interpositions w, x, y, z, then the marked positions on the string between 1 and 2 (an octave) and between 2 and 4 (also an octave), would read as follows:

$$
\begin{array}{ccccccc}
 & & & (9/8) & & & \\
1 & w & x & \mathbf{4/3} & 3/2 & y & z & 2 \\
2 & 2w & 2x & \mathbf{8/3} & 3 & 2y & 2z & 4 \\
\end{array}
$$

Let us recall Plato's text concerning this latter procedure:

> He cut off yet further portions from the original mixture, and set them in between the portions above rehearsed, so as to place two Means in each interval—one a harmonic Mean; the other an arithmetical Mean. And whereas the insertion of these links formed fresh intervals in the former intervals, that is to say, intervals of 3:2 and 4:3 and 9:8, He went on to fill up the 4:3 intervals with 9:8 intervals. This still left over in each case a fraction, which is represented by the terms of the numerical ratio 256:243.

This passage has been a crux for a great many interpreters. Indeed between 1 and the harmonic Mean 4/3, one observes a string interval of ratio 4/3. This interval must be filled up with 9/8 ratios (with 9/8 or a multiple thereof) since the latter gives a whole tone in Greek music.[6] This boils down to a twofold insertion, viz., 9/8 and 81/64 (or 9/8 in the second power); w and

6   Ibid., 140.

x are thus identified as 9/8 and 81/64, respectively. This ascending progression indicates the lengthening of string sections, and consequently, a progressive lowering of the tones; if one plucks the longer section between 1 and 81/64, one gets a sound that is a whole tone lower than that resulting from the plucking of the string section between 1 and 9/8. The same is true if one plucks the string section between 1 and 4/3, but in this case—as we shall see—the lowering of the sound turns out to be less than a whole tone. Let us term this: re, do, ti.

As regards the string interval between the arithmetical Mean 3/2 and the integer 2, the same procedure must be applied. This string segment has a ratio of 3/2, so between its extremes, two insertions of ratio 9/8 can be interposed, viz., 27/16 (or 3/2 x 9/8) and 243/128 (or 3/2 x 9/8 x 9/8), which again gives two whole tones. If one plucks the string section between 1 and 243/128, the outcome is also a note that is a whole tone lower than that produced by the plucking of the string section between 1 and 27/16. Compared to this, the string section between 1 and 2 gives a sound that is again lower than a whole tone. Let us term these three tones sol, fa, mi. If we add to this the whole tone that results from the ratio 9/8 between the harmonic Mean 4/3 and the arithmetical Mean 3/2, and we term this note la, we get the full octave: re, do, ti, **la**, sol, fa, mi, (re).

1	9/8	81/64	**4/3**	3/2	27/16	243/128	2
a	a	b	a	a	a	b	
re	do	ti	**la**	sol	fa	mi	

In this octave (the Phrygian mode), there are two tones smaller than a whole tone; the two have been represented by b. These smaller tones—they only approximate standard European semitones—were termed in Plato's text a *leimma* or rest; their ratio is 256/243, which is not reducible to 9/8 or a multiple thereof. What Plato says about these fractions can be checked. The string section between 243/128 and 2 is characterized by the ratio 2/1: 243/128, or 256/243 (indeed, 2 x 128 = 256, and 1 x 243 = 243), and so too, the string section between 81/64 and 4/3 shows a ratio of 4/3: 81/64, or 256/243 again (indeed, 4 x 64 = 256, and 3 x 81 = 243). One ought not to forget that the Greek semitone is shorter than the standard European tone; the Greek semitone is between a semitone and a fourth of the European one. On the other hand, their whole tones were a bit longer than the European one. Plato would have experienced later Western music as strange and cacophonous.

The "bridging" of all the intervals between 1 and 27 yields the following result:

1	9/8	81/64	4/3	3/2	27/16	243/128	2
a	a	b	a	a	a	b	
re	do	ti	la	sol	fa	mi	
(2)	9/4	81/32	8/3	3	27/8	243/64	4
a	a	b	a	a	a	b	
re	do	ti	la	sol	fa	mi	
(4)	9/2	81/16	16/3	6	27/4	243/32	8
a	a	b	a	a	a	b	
re	do	ti	la	sol	fa	mi	
(8)	9	81/8	32/3	12	27/2	243/16	16
a	a	b	a	a	a	b	
re	do	ti	la	sol	fa	mi	
(16)	18	81/4	64/3	24	27		
a	a	b	a	a			
re	do	ti	la	sol			

Figure 2. Musical scales: descending melody

The descending series of gamuts ending in a deep-sounding bass sol is longer than what a Greek instrument was able to play, but it ought to be that long in order to evoke the music of the spheres. Because the whole series is proportionally bound together, it can be played both forward and backward. In the latter case, it is as if, from the depths of the universe, deep-sounding bass notes are rolling to us, becoming higher and higher the more they reach the near planets. Conversely, the high-sounding notes emitted by the near planets can be regarded as the beginning of a series of gamuts whose descending heavier notes roll farther into the more remote parts of the universe.

Intermezzo. The World Soul and the Self-activation of the Universe: Whitehead's Appreciation of Plato

The reader may perhaps ask why Plato is so eager to present a detailed picture of the intrinsic harmonious proportions of the World Soul. The answer is obvious: He seeks to show that the World Soul, being the vivifying principle mediating between ideal forms and material world, is simply the most perfect and solid architectonic construction the Demiurge has undertaken to create. Not only is the World Soul beautiful, she is also extremely solid. Whenever an interval is filled with a fresh ratio, its two extreme terms are being glued together by means of precise links. Indeed, all the numbers of the series of the powers of 2 and the powers of 3 are tied together by segments of the numerical ratio 9/8 and 256/243; this provides the World Soul with a firmly knotted nervous system, so to speak, from which she draws her quasi inde-structible energy for immanently acting upon the world. This aspect cannot be underlined enough. The World Soul unfolds her indestructible energy in the material world in virtue of being held together by multiple chains of golden proportions. This is the message Plato addresses to the contemporary reader with his theory of the mathematical and harmonic Means.

Plato confronts us with a cosmos in the heart of which God has placed a varied and solid impetus toward self-preservation. To account for this immanent self-sustaining force, Plato has spared no efforts to describe the marvelous composition of the energetic World Soul. Only a perfectly propor-tionate World Soul is capable of animating the world as it should be.

That is the reason why Alfred North Whitehead (1861–1947), the orig-inator of Process Cosmology, esteems Plato so much. For him, Plato offers a welcome complement to the Hebrew notion of an externally creating God. Plato makes us familiar with a God who places in the very cosmos all the necessary conditions required for its amazing self-preservation (and Whitehead would add, for its striving towards novelty). Indeed, whatever is tied together by virtue of proportionate ratios is able to sustain and renew itself accordingly. The *Timaeus* explicitly explores the perspective of an immanently organic unfolding of the cosmos, made possible by the golden proportions the Demiurge has so carefully placed in the World Soul. This immanent unfolding allows the various entities in the cosmos to be sensitive to the entry of ever new forms and combinations into their real potentialities. According to Whitehead, Plato laid down the foundations of the "evolution of matter" towards ever new clusters of order. This is in sharp contrast to Newton, who in the context of a totally mechanical view of the world, opted for the existence of stable matter, not capable of organizing itself in terms of ever new combinations.

For Whitehead, Plato was ahead of his time for two reasons: (a) Plato took seriously the sharp-cut differences between the basic elements in nature by bringing them into a rapprochement to the forms of regular solids: pyramid, octahedron, icosahedron, and cube: "While we note the many things said by Plato in the *Timaeus* which are now foolishness we must also give him credit for that aspect of his teaching in which he was two thousand years ahead of his time. Plato accounted for the sharp-cut differences between kinds of natural things, by assuming an approximation of the molecules of the fundamental kinds respectively to the mathematical forms of the regular solids."[7] (b) Plato was able to explain qualitative contrasts, such as those between musical notes by means of the insertion of simpler ratios between integral numbers. This proliferation of simpler ratios, however, did not lead to the formation of pure dissonances, since all the ratios were harmoniously linked together by mathematical and harmonic Means: "He, thus, obtained a reason why [...] there should be sharp-cut relations of harmony standing out amid dissonance." Because of these achievements Plato, in a sense, anticipated the breakthrough of quantum physics in the 1920s: "Newton would have been surprised at the modern quantum theory and at the dissolution of quanta into vibrations. Plato would have expected it."[8]

In his ground-breaking work *Process and Reality*, Whitehead integrated the basic insights of quantum physics. This allowed him to develop his organic philosophy with its focus on the ever new emergence of unexpected constellations. When looking back at Plato's *Timaeus*, he acknowledges that this work already contained the germs of the evolution of matter. He writes: "There is another point in which the organic philosophy only repeats Plato. In the *Timaeus* the origin of the present cosmic epoch is traced back to an aboriginal disorder, chaotic according to our ideals. This is the evolutionary doctrine of the philosophy of organism."[9]

In his assessment of Plato, Whitehead contrasts the latter with Isaac Newton (1643–1727), who in line with the Jewish tradition upheld the total transcendence of the God-Creator. In this view no place can be given to the vision Plato elaborated with his notion of the World Soul, namely the organic unfolding of the cosmos, or in Whitehead's terminology: the evolution of matter:

> Plato's notion [of an aboriginal disorder] has puzzled critics who are obsessed with
> the Semitic theory of a wholly transcendent God creating out of nothing an accidental
> universe. Newton held the Semitic theory. His *scholium* made no provision for the
> evolution of matter—very naturally since the topic lay outside its scope. The result
> has been that the non-evolution of matter has been a tacit presupposition throughout
> modern thought. Until the last few years the sole alternatives were: either the mate-
> rial universe, with its present type of order, is eternal; or else it came into being, and
> will pass out of being, according to the fiat of Jehovah. Thus, on all sides, Plato's

---

7   Alfred North Whitehead, *Process and Reality. An Essay in Cosmology*, Gifford Lectures 1927–
     1928 (Toronto: Macmillan, 1927 [1969]), 113.

8   Ibid.

9   Ibid., 114.

allegory of the evolution of a new type of order [...] became a daydream puzzling the commentators [...] Both for Plato and for Aristotle the process of the actual world has been conceived as a real incoming of forms into real potentiality [...]. Also, for the *Timaeus* the creation of the world is the incoming of a type of order establishing a cosmic epoch. It is not the beginning of matter in fact, but the incoming of a certain type of [...] order.[10]

Cosmic Ballet and the Music of the Spheres

Proportional ratios also play an important part in Plato's astrological considerations. Here too, he is indebted to Pythagoras, who used to combine delight in mathematics with contemplation of the heavenly ballet resulting from the changing positions of stars and planets. Because they lacked telescopes, ancient astrologers had to study the astral phenomena with the naked eye. The only technical means at their disposal were astrolabes, which allowed them to measure the angle under which one could see the Moon, Mercury, etc. relative to the fixed stars, and to calculate their changing positions. With the help of this technique, Greek astrologers were able to surpass their Babylonian colleagues.[11]

Astrologers were present in all the great ancient civilizations: Inca, Chinese, Egyptian, and Babylonian. No neat separation existed between astrology and astronomy. Extraordinary astral phenomena like shooting stars were interpreted as announcing unexpected events. Astronomy also had political implications because it allowed one to fix or to readjust the usual calendar in the name of some ruler. Instead of the name of astronomers, that of the reigning emperor was being added to the new calendar system:

> In Egypt, ever since the Ancient Empire, the calendar was a public affair. It is the name of Julius Caesar, not of his astronomer Sosigenes, that figured on the calendar which dominated the whole history of Europe until the end of the 16th century. The revised calendar under which we live now bears the name of a pope, Gregory XIII, and not of his astronomer.[12]

More than to the oracle function of stars and planets, Plato pays attention, as we will see, to the harmonious proportions between the respective orbits of the planets. From the astronomy of his day, he adopted the distinction

---

10   Ibid. 114–15.

11   See Rom Harré, *The Philosophy of Science* (Oxford: Oxford University Press, 1984), 47.

12   Jacques Merleau-Ponty and Bruno Morando, *Les trois étappes de la cosmologie* (Paris: Laffont, 1971), 38 (translation mine).

between the fixed stars (such as the Polar Star, the Big Dipper, the little Dipper, and Orion) and the planets (the Sun, the Moon, Mercury, Venus, etc.). *Planein*, in Greek, means "to wander." So, "planets" was the name given to stars that seemed to move from night to night, when measured against the self-same position of the fixed stars.

It was believed that the Earth was in the center of the universe and that the fixed stars as well as Sun, Moon, and the other planets moved around the Earth. According to ancient cosmology, the universe is spherical. Like every sphere, it has a centre and an axis. The point directly above an observer on earth is termed the North Celestial Pole, whereas the point directly beneath that observer is the South Celestial Pole. Between both extremes is the place of the Earth, which in spite of lacking a proper rotation, is thought to be slightly rocking.

Perceived from the Earth, the firmament with the fixed stars seems to carry out a daily rotation of 360 degrees. Each night one can see the starry sky rise to set again at dawn, and the scenario starts over again the following night. This rotation takes place from the east to the west (after the Sun has set in the west the starry sky will rise again in the east). As far as the planets are concerned, they share in this westbound rotation, but at the same time, they seem to possess in addition a proper movement that carries them from

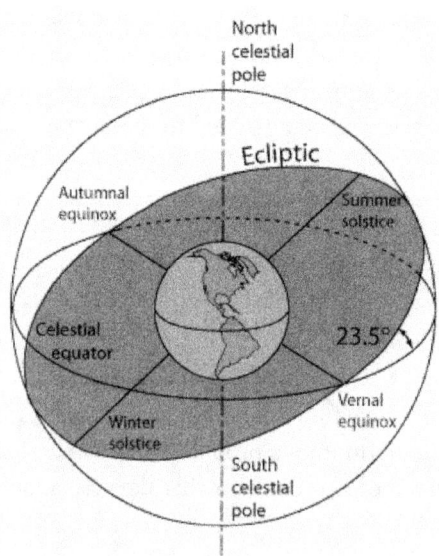

Figure 3. Path along the Ecliptic

Source: Hyperphysics, Georgia State University
(http://hyperphysics.phy-astr.gsu.edu/hbase/eclip.html), accessed September 19, 2014.
This figure is reproduced courtesy of Professor Carl R. Nave, Georgia State University.

the west to the east; this makes them gradually lag behind the westbound rotation of the fixed stars. This peculiar "recession," moreover, follows a circular path whose plane is tilted at an angle of approximately 23.5° to the plane of the equator of the universe.

This path is termed the ecliptic because on it such rare phenomena as solar eclipses take place. Indeed, the Sun is observed to travel along this apparent path. This journey has a bearing on the seasons. Their rhythm in the Northern hemisphere reads as follows: At the Vernal equinox, on March 22, when the Sun passes from the Southern to the Northern hemisphere at a point that intersects the celestial equator, the day is as long as the night; the lengthening of the days commences. At the Summer solstice, June 21, the Sun is farthest north; so, the daytime is the longest of the year. At the Autumnal Equinox, September 22, the Sun crosses the equator moving southward; this inaugurates the beginning of the shortening of the days. At the Winter solstice, December 22, the Sun is farthest south; so the daytime is the shortest of the year. In the Southern hemisphere the rhythm of the seasons follows the opposite scheme: Winter in the Northern hemisphere means summer in the Southern hemisphere, etc.

In Ancient Greece, New Year began with the Spring Equinox on March 22. This date was taken as reference point from where to divide the ecliptic into 12 equal segments, corresponding to the 12 months. At the Spring Equinox the Sun was in the constellation Aries, and from there began to traverse the subsequent constellations of the Zodiac: Taurus, Gemini, Cancer, Leo, Virgo, Libra, Scorpio, Sagittarius, Capricorn, Aquarius, and Pisces. The Sun traverses the full length of the Zodiac once each year.

More particularly the ancients observed that all the planets, each with its own rhythm, moved eastward through the Zodiac along the oblique circle of the ecliptic. It takes the Moon only a month to traverse its path along the ecliptic, whereas for the Sun this travel takes 365 days. For Mercury and Venus this revolution is shorter than a year, but for the "outer planets" it takes years to traverse the path along the ecliptic: 2 years for Mars, 12 years for Jupiter, and 30 years for Saturn.

When Plato, in the *Timaeus*, sets out to explain the layout of the starry sky, he gives evidence of perfectly knowing the difference between the fixed stars and the planets. In order to describe the respective distance of the planets from the earth, and their respective speed, he has recourse to proportional ratios, just as he had done in his elucidation of the construction of the World Soul. Hence most of the commentators believe that, for Plato, the starry sky is simply the World Soul, or at least, its most visible appearance. Plato conceived the universe as consisting of two large concentric circles: one

steered by the rhythm of the "same," and the second by that of the "other."
He writes:

> Next He [the Demiurge] split all this that He had put together [the World Soul]
> into two parts lengthwise; and then He laid the twain one against the other, the
> middle of one to the middle of the other, like a great cross; and bent either of them
> into a circle [...] And he compassed them about with the motion that revolves in
> the same spot continually, and He made the one circle outer and the other inner.
> And the outer motion He ordained to be the Motion of the Same, and the inner
> motion the Motion of the Other. And he made the Motion of the Same to be toward
> the right along the side, and the Motion of the Other to be toward the left along
> the diagonal. (*Tim.* 36cd)

That is, along the ecliptic.

In a next sequence, Plato narrates how the Demiurge divides the inner
circle into the respective circles of the seven planets:

> He split the inner Revolution in six places into seven unequal circles, according
> to each of the intervals of the double and triple intervals, three double [2,4,8] and
> three triple [3,9,27]. He appointed three [the orbits of Sun, Mercury, and Venus]
> to revolve at an equal speed, the other four [the orbits of Moon, Mars, Jupiter, and
> Saturn] to go at speeds equal neither with each other nor with the speed of the
> aforesaid three, yet moving at speeds the ratios of which one to another are those
> of natural integers. (*Tim.* 36d)

What is important for Plato is that, in spite of their peculiar journeys,
the planets traverse tracks that are at a proportionate distance both from
one another and from the centre of the Earth, for only a bond as solid
as that established by means proportional can really tie them together.
When the distance of the Moon to the Earth constitutes a first unit (1),
then the Sun ought to be at a double distance (2), Mercury three times that
far (3), Venus four times (4), Mars eight times (8), Jupiter nine times (9),
and Saturn 27 times that far (27). In this calculation, one easily recognizes
the combined series of the powers of 2 and the powers of 3, typical of the
structure of the World Soul: 1, **2**, 3, **4**, 9, **8**, 27. On this proportional series
rests the imperturbable common harmony among the orbiting planets in
spite of the fact that their respective rotations deviate from that of the
firmament. One will have to wait till Ptolemy (2nd c. AD) before the Sun
will be given a higher place in the series of the planets, a place between
Venus and Mars.

Apart from their specific speed, all the planets, on their route along the tilted plane of the ecliptic, display a varied ballet of perfect circular movements around the centre of the universe. They go on repeating this varied ballet against the background of the westbound rotation of the firmament. It is this grouping of regular and proportional circularities that in the last resort accounts for the music of the spheres.

With his contemporaries, Plato held in great esteem the motif of the circular trajectories of stars and planets (the idea of elliptical structures would not have come to their minds). The celestial bodies constituted an object of worship because of their most perfect circular orbits. To contemplate in the sky these gracefully rotating forms is to experience eternity in the midst of temporality. For Plato, the periodic recurrence of the astral formations is "a movable image of Eternity" (*Tim.* 37d). He is so deeply impressed by astral religiosity that he yearns for the advent of the "great cosmic year" in which the wandering planets will regain their most harmonious positions. In the "great cosmic year," the ancient world celebrated the periodically returning grand cosmic order that must have existed "in the beginning of creation."

The Creation of Animals and Humans

Plato does not eschew mythological language. He attributes a divine dignity to the stars and planets as if they were deities in their own right ("the Gods who revolve manifestly" [Tim. 41a]). The ancient Greeks used to worship, besides Zeus (Jupiter), minor heavenly gods. In Plato's narrative they are appointed by the Demiurge to be his co-creators (secondary causes). Yet, before giving them this mission, the Demiurge reminds them of their being created by Him in their quasi-immortal existence ("Wherefore ye also, seeing that ye were generated, are not wholly immortal or indissoluble, yet in no wise shall ye be dissolved nor incur the doom of death" [*Tim.* 41b]). The minor gods' task is to fashion the living creatures: the plants, the animals and even the humans, for it is said, if the Demiurge himself would directly shape the humans, they would enjoy some immortal mode of existence, and that would not be proper:

> If by my doing these creatures came into existence and partook of life, they would be made equal unto gods; in order, therefore, that they may be mortal and that this World-all may be truly All, do ye turn yourselves, as Nature directs, to the work of fashioning these living creatures, imitating the power showed by me in my generating of you. Now so much of them as it is proper to designate "immortal," the part we call divine which rules supreme in those who are fain to follow justice always

and yourselves, that part I will deliver unto you when I have sown it and given it origin. For the rest, do ye weave together the mortal with the immortal, and thereby fashion and generate living creatures, and give them food that they may grow, and when they waste away receive them to yourselves again. (*Tim.* 41cd)

Though conveyed in mythological language, Plato's message is unambiguous: it is not unworthy of God to let other forces ("minor deities") share in his creative power; God dares to empower secondary causes to be creative agents in the universe. The Demiurge charges the minor gods to fashion living creatures, to form their biological components and their animal soul, and, as far as the humans are concerned, to unite their biological organism to the immortal part of the soul (the *nous* or spirit) that God alone is going to create. This division of labor between God and the second causes would be inconceivable in the Semitic life world. In Genesis, God alone creates the various species, not in the least assisted by any minor deity.

In the second biblical creation narrative, too, special attention is given to the creation of the human soul, i.e. the human life principle that includes the intellect. JHWH "breathed into his (Adam's) nostrils the breath of life." This happened after God Himself had "formed Adam from the dust of the ground" (Gen. 2:7). God alone creates body and soul. Yet, another difference comes to the fore. In the *Timaeus*, the mortality of the body is taken for granted; this is mythically suggested by the Demiurge's refusal to directly fashion the human bodies, for then they would be immortal. In the Bible, on the contrary, humankind's mortality is presented as the seal of sin, as the consequence of some mysterious disobedience in the Garden of Eden. Adam and Eve have been seduced by a serpent who tells them: "Of course, you will not die. God knows that as soon as you eat it [that fruit], your eyes will be opened and you will be like gods, knowing good and evil" (Gen. 3:2). Monotheism apparently needs human sin to account for the mortal condition of the human species. A remarkable fact is that Catholic theology tends to resuscitate the Platonic narrative about a God who reserves to himself the task of creating the human soul. In November 1996 Pope John-Paul II sent a message to the Pontifical Academy of Sciences, in which he basically recognized the merits of the theory of evolution, as proposed by Darwin and others. Yet, in spite of this openness to the theory of evolution John-Paul II reiterated the note of caution Pius XII already advanced in 1950: even if one admits that the human body takes its origin from pre-existent living matter, the spiritual soul is immediately created by God.

CHAPTER 3

Aristotle

After Plato's death in 347 BC, his school became divided over the question as to how to faithfully continue their master's thinking. The victory was won by those adepts who were in favor of a purely mathematical explanation of his legacy: the Platonic ideas are best rendered by mathematical and geometrical proportions. Aristotle (384–22) disagreed; he left the academy to develop a method that focused on the physical growth of things.

Physics instead of Mathematics

For Aristotle, the Platonic ideas ought to be regarded as active forms operative within material beings themselves. Pure mathematics fails to see this aspect of growth and development. Aristotle's point of departure is the intrinsic connection between form (*morphe*, *forma*) and matter (*hyle*, *materia*). Each entity is endowed with a formative principle that steers the unfolding of its material component. In terms of "potentiality" (*potentia*) and "actuality" *(actus)* this means that the potential stuff grows toward its fulfillment because it is moved to it by the kinetic force of its substantial form. Potentiality turns into complete realization (*actus)* under the impact of its steering form that dictates the end.

Aristotle's work is regarded as a classic contribution to the fields of ethics, politics, physics, and metaphysics. What is perhaps less known is that he was also engaged in the study of zoology and the classification of living things, as can be seen in his works *History of Animals*, and *Generation of Animals.* No wonder, thus, that the root metaphor he used to study various domains of life was that of biological development:

"One of the main sources of Aristotle's thinking was the observation of embryonic growth."[1]

Aristotle sought to get an insight into the mechanisms that structure the development of life on earth. For him, such a comprehensive view ought to begin with the study of the motions to be found in nature. Life on earth is characterized by motion and becoming; yet, in order to account for the ever-recurring rhythm of origination, growth, and perishing in the sublunary sphere, one must have recourse to the regular circular rotations of the celestial bodies. These form, so to speak, the "immortal" background against which life on Earth, from generation to generation, is going to acquire stable patterns in spite of incessant change. Like most of his contemporaries, Aristotle was fascinated by astral religiosity.[2] This led him, as we will see, to the recognition of a prime mover: the first cause of the regular motions of stars and planets, and of the recurrent patterns of order on Earth. Hence his famous dictum: "if a thing is in motion, it is, of necessity, being kept in motion by something else" (*Physics*, VII, I, 241b 24).[3]

It is useless to look for a creation narrative in Aristotle's writings. Not interested in a creation at the beginning, he rather takes delight in the basic patterns of order that perpetuate themselves in all that moves and lives. In that sense, he represents the emergence of the "scientific mind," which replaces mythical sayings about the origin of the universe with the study of real causes. Of course, the connotation of "scientific" must be nuanced. Although Aristotle's *Physics* remained in vogue for almost 2,000 years, its basic tenets can by no means stand in comparison with modern physics (Galileo, Newton). I give a few examples.

(i) Whereas modern mechanics thinks in terms of the interaction of bodies on the basis of their mass, Aristotle neglected the importance of interactions. His basic model is that of living organisms that strive to attain the perfection of their intrinsic form.

(ii) Aristotle starts from the assumption that a body, when left to itself, must be at rest unless its rest is being disturbed by something else. Modern mechanics has corrected this view and demonstrated that the law of inertia (the tendency of a body to remain in its original state) not only applies to

---

1   Ilya Prigogine and Isabelle Stengers, *Order out of Chaos: Man's New Dialogue with Nature* (London: Fontana Paperbacks, 1984 [1986]), 40.

2   Pierre Aubenque, *Le problème de l'Etre chez Aristote* (Paris: Presses Universitaires de France, 1962), 339.

3   Aristotle, *The Physics*, II, ed. Philip Wicksteed and Francis Cornford, The Loeb Classical Library (London: Heinemann, 1968).

bodies at rest but also to bodies in motion. In Newton's formulation: "Every body continues in its state of rest or of uniform motion in a right line unless it is compelled to change that state by forces impressed on it."[4] Aristotle also opined that heavier bodies fell faster towards the Earth than lighter ones, a view Galileo (1564–1642) was later to refute with his experiments.

(iii) Aristotle's reservations about the use of mathematics historically blocked the avenue for the achievements of the modern natural sciences. Physical reactions cannot possibly be predicted, unless experimental results are couched in mathematical equations.[5] Aristotle's method is, in the first instance, descriptive and classificatory rather than truly explanatory.

Nature in Motion

Aristotle discerns two basic categories of motion: natural motion and violent motion. Locomotion in space, e.g., can be a natural motion (as is the case with the fall movement), but can also be a violent motion, such as the throw of a lance or a discus or the traction exerted on a ship to drag it out of the sea.

(a) Natural motion. A first type of natural motion is that resulting from a substantial form that acts upon its material component. The "something else" that moves the entity is, in this case, the substantial form. The human body is moved by its soul, i.e., by its formative principle. The same is true for animals, plants, and inorganic entities; they are all moved and vivified by the kinetic force of their substantial forms. This kinetic force does not proceed from an external mover, but from an intrinsic mover. The substantial form steers the object's motion from within.

Underlying this view is Aristotle's notion of *entelechia* or intrinsic finality. Unlike Plato who placed the ideal form outside matter, Aristotle regards the form (*morphe*) as the intrinsic principle of organization within a concrete thing. It is "from within" that the form makes matter (*hyle*) strive to acquire a particular configuration. Plants, animals, and human beings all possess an appetite for growth because of the finality contained in the formative principle. Besides this, Aristotle also draws a distinction between the substantial form and the accidental form. Although an accidental form may add some variety, only the substantial form determines the basic development of an organism. It is more vital to be a mature human person than to be a white or

---

4   Isaac Newton, *Principia Mathematica*, ed. CAJORI , Vol. I, *The Motion of Bodies* (Berkeley and Los Angeles: University of California Press, 1966), 13

5   Jacques Merleau-Ponty and Bruno Morando, *Les trois étappes de la cosmologie* (Paris: Laffont, 1971), 97.

a black person, the latter quality only being a variation in the human species. The same is true for plants and animals: variations among the same species flow from the working of the accidental form.

In order to further clarify the natural motion of things, Aristotle distinguishes four causes: material cause (what matter is involved?), formal cause (what is the inner formative principle?), efficient cause (what is the entity bringing about the change?), and final cause (what end is being pursued?). These aspects are interrelated. The formal cause practically coincides with the final cause, whereas formal and material causes seem to presuppose each other. The efficient cause consists in the kinetic power that produces the effect.

Under the rubric "natural motion" also come various types of locomotion that occur in the natural setting of the geocentric universe. Locomotion takes place in the "chemical reactions" between the basic elements fire, air, water, and earth. Each basic element possesses a primordial quality and an accessory quality that, in turn, can be a primordial quality in another element. This makes mergers possible. Earth is dry and cold; water cold and humid, air humid and warm, and fire warm and dry. When massive bodies made up of earth (dry and cold) are bonded with water (cold and humid), liquefaction occurs, such as the melting of ores. So, too, when fire (warm and dry) and earth (dry and cold) combine, smoke arises. In all these cases displacement of qualities occurs.[6]

Locomotion also occurs in the ascending or descending motions of the four elements in free space. Fire naturally tends to move upward, whereas earth tends to fall down (one ought only to drop a lump of earth, and it falls down). This gives us the two basic types of locomotion: the rectilinear movement upward and the rectilinear movement downward. Water and air display a rather referential behavior. Relative to air, water falls down (in the form of rain), whereas relative to earth it rises (the swelling of rivers after rainfall). Water and air seem to act as buffers in exerting a resistance against the rectilinear movements upward and downward, both of which they curb. Freed of any restriction, however, fire moves "upward" thanks to its accidental quality "lightness" (*leve*), whereas earth moves "downward" thanks to its accidental quality "heaviness" (*grave*). In this way they both tend to rush towards their "natural place": the region of the moon's orbit, for fire, and the Earth in the center of the universe, for the element earth. At this juncture Aristotle observes that in both cases the rectilinear motion is finite. Perpetual motion is apparently not part of our globe: it is the sole privilege of the

---

6   Ibid., 22–23.

cosmic regions in which stars and planets orbit along circular paths without beginning or end. The celestial rotating hoops in which stars and planets are set traverse their paths after the perfect model of the circle.[7]

(b) In the sublunary sphere all the motions are finite in character. This is also true for the category of violent motions. They, too, have a beginning and an end. The classic example is the throw. I dwell on it because the reasoning used, viz. transmission of the kinetic force through intermediary movers in close contact with each other, will return in Aristotle's proof of God's existence in his book *Physics* (see below). What happens when one flings a stone? The projectile dashes forward before eventually reducing speed and falling down. Aristotle attempts to explain what happened in the meantime. In casting a stone, the "thrower" gets it going, but also passes on an amount of kinetic force to the first layer of air that is going to carry the stone, and from this layer of air the kinetic force will be passed on to the next layer of air. This leads to a linked series of adjacent "intermediary movers" (*motores conjuncti*), each of them passing on the kinetic force to a next "intermediary mover," till the stone begins to fall. The person who throws the stone is initially in close contact with it and, thus, functions as the "prime mover" (*primus movens*). He passes on the kinetic force not only to the projectile but also to the adjacent layer of air, which acquires the capacity to pass it on to the next adjacent layer of air, which in turn…, etc. In this way, the projectile meets at each point of its path an "intermediary mover" required for the maintenance of the motion, while the motion itself in the course of the transmission gradually lessens its force till the moment comes when a further adjacent layer of air no longer gets enough kinetic force to pass the motion on any further. At that moment the projectile takes on its natural fall movement.[8]

The above example calls to mind Aristotle's basic rule: "nothing is in motion unless it is moved by something else." As a matter of fact, we have to do with a linked series of "movers" in which the kinetic force is passed on from adjacent intermediary mover to adjacent intermediary mover. Yet, for this force to be properly transmitted through the whole series, close contact between the first mover and all the adjacent intermediary movers is needed. Without this close contact the initial kinetic force cannot reach the last link in the concatenation. This is in sharp contrast to Newton's later concept of "action at a distance." For Newton, as we will see, the force of gravity pierces through the void; it makes its impact felt on objects at a distance, without any close physical contact.

---

7   Ibid., 25.
8   Ibid., 27–28.

Cosmic Movers "Up There" and the Quest for a Prime Mover

Looking at the starry sky, one observes how stars and planets traverse their circular paths without having to overcome any resistance. From this, Aristotle deduces that the motions "up there" are of a different order than the natural and violent motions on Earth. The undisturbed regularity of the rotations of the rings or spheres in which the stars and planets are set cannot be explained by some kind of vacuum. Such would imply the absence of loco-motion. So, the starry sky must be made of a special kind of matter, called quintessence, the fifth element besides fire, air, water, and earth. All things partaking in it—the firmament, the planetary rings as well as the stars and planets themselves—are regulated by a "movement without beginning and without end" (*De Coelo*, 270 b, 20)[9]: they are everlasting and indestructible, in contrast to the entities and motions in the sublunary domain, which are all finite in nature.

Let us recall that in Plato's *Timaeus* stars and planets were called "the Gods who revolve manifestly" (*Tim.* 41a). And also that these "heavenly Gods" acted as "secondary causes" in creation: their task was to "fashion the bodies of animals and humans," and to "give them food that they may grow." Aristotle abhors mythical explanations. For that reason he has no creation story. Yet, in spite of this reservation, he has something in common with Plato. He takes it for granted that the heavenly quintessential realm exerts an influence on life on Earth. For him, the never-ending regular rotations of the heavenly circles produce the kinetic force that steers the rhythms of development and growth in the sublunary sphere. It is thanks to the kinetic thrusts that spurt down from heaven that the varied kinds of organisms enact their transition from potentiality (*potentia*) to actuality (*actus*), after which they begin to fade away.

This is for Aristotle an occasion to engage in a reflection on the begetting (*generatio*) and decease (*corruptio*) of organisms. In the course of the succession of generations, individual members of the species pass away. Yet, the species as such continues to exist because, in the act of procreation, the begetters pass on their "genetic code" (the formative principle of the species) to the future offspring. It is this "genetic code" (form, *morphe*) that secures the continuation of the respective species in spite of the mortality of their single members. Aristotle is fascinated by this ever recurring continuity: he sees it occur in the enormous variety of plants and animals, andin the human species at large. For him, this ever recurring reproduction of the human species can

---

9   Aristotle, *On the Heavens*, ed. W.Guthrie, The Loeb Classical Libary (London: Heinemann, 1970).

only be explained by the never-ending heavenly motions "up there." The seeds of humanity remain mortal; it is only from the permanent rotations of the stars and planets that a promise of perpetuity is given. Behind the father fathering a child, one ought to imagine the thrusts of the heavenly spheres, whose vivifying power breathes life into the newly conceived offspring of the human species. This is encapsulated in Aristotle's dictum, reiterated by Thomas Aquinas in the 12th century: "The human being is fathered by his father and by the Sun" (*Physics*, II, ii, 194b 13). Without the assistance of the Sun reproduction as it should be—with its promise of perpetuity of the species—cannot occur. What is true for humankind is also analogously true for the other living species on Earth.

The rhythms of origination, growth, and perishing make Aristotle, thus, look for cosmic movers capable of fostering the entry of new life into the perishable earthly tissue. He admits, in the wake of Plato, the existence of a vitalizing cosmic World Soul, whose differentiated host of celestial "movers" warrants the renewal of life on Earth. He even dares to name the core of this World Soul "the highest God," who needs nothing but his own proper motion to exist (*autokinetos*). This God passes on the thrust of his kinetic force to the other heavenly regions, to the rotating firmament (*primum mobile*) first, and from there to the rotating rings of the seven planets till the transmitted kinetic force reaches the Sun, the Moon, and finally, the sublunary realm.

Aristotle ascribes a special dignity to the Prime Mover, the highest God on whose primordial motion all the derivative forms of cosmic motions depend. The Prime Mover's influence extends over the whole domain of the spherical universe; his dwelling place is located above the firmament, whose rotation he directly sets in motion, whereas the rotating firmament communicates this motion to the planetary spheres. Through the medium of a differentiated clock-work of interposed movers, God's energetic force stretches down to the Earth to provide it with a glimpse of durability in spite of the counter-rhythms of perishing. The Prime Mover is the highest God on top of the emanating chain of cosmic movers.

The above description presupposes, and relies on, the proof of God's existence Aristotle gives in his *Physics*. To the presentation of this proof I now turn. The proof has two parts. In the first part Aristotle demonstrates (a) the existence of cosmic movers and in the second part he demonstrates (b) the existence of the Prime Mover.

(a) With his contemporaries Aristotle shares the basic conviction that the heavenly ballet of stars and planets discloses what eternity and immortality means. "Up there" is the reign of the "Cosmic God,"[10] whereas life on Earth is characterized by impermanence and finitude. So, when indications of durability are to be perceived on Earth, they must have their origin in this heavenly region. Aristotle gives a clearer articulation to this postulate by showing that the transition from potentiality to actuality, as well as the transmission of the "genetic code" from generation to generation, cannot be effected without appealing to cosmic movers. These provide in the last resort the kinetic energy that is required to bring about these processes.

(b) In order to demonstrate the existence of the Prime Mover, Aristotle establishes a hierarchical order among the various cosmic movers in heaven. The argument takes on the style of pure logic; yet, what is at stake is the dignity of the "highest God," the Prime Mover. He writes:

> The thing moved may be moved by the true agent either directly or by some inter-mediate which itself is moved by the true agent directly. And the true agent may immediately precede the intermediate agent which acts directly upon the patient, or there may be a chain of several intermediates. (*Physics*, VIII, v, 256a 4–6)

Aristotle clearly opts for a chain of several intermediates. In the concrete, this means that the "highest God" (the true agent) directly sets the firmament (the first intermediate agent) in motion, which, in turn, also through direct contact—sets the sphere of Saturn (the second intermediate) in motion, which in turn through direct contact sets the sphere of Jupiter (the third intermediate) in motion, etc. . . . till in this chain of intermediates the sphere of the Moon is set in motion, which directly acts upon the enti-ties and the organisms on Earth. Aristotle thus applies the schema he used in his explanation of the throw (the violent motion) to the transmission of motion from the Prime mover up to the sphere of the Moon. Again, the kinetic force originating from the "true agent" (the "true mover") is passed on to all the intermediate layers. This time, however, since we are no longer in the realm of finite motions, this transmission takes place without any diminution of the initial kinetic force. It is important to note that the transmission of kinetic power from agent to agent takes place through direct contact.

Having given this explanation, Aristotle looks again at the chain of intermediaries, this time from the vantage point of the lowest intermediate agent. From this vantage point one only perceives an ascending succession of intermediate agents. Yet, Aristotle will argue, an infinite regress without

---

10   André-Joseph Festugière, *Le Dieu Cosmique* (Paris: Les belles letters, 1949).

reaching a true, independent initiator is impossible. At the beginning of the chain there must be a self-moving mover. He writes:

> However long the chain, therefore, of things that produce motion by an instrumentality other than their own, there must lie behind it an agent that produces the movement by its own instrumentality. So that if this primary agent is in motion, and there is no agent behind it to set it in motion, it must of necessity be moving itself. So this line of argument again leads to the conclusion that if anything is in motion it must either be set in motion by a self-moving agent immediately, or must send us back through a chain of intermediaries until we come to such an agent. (*Physics*, VIII, v, 256a 31–b3)

In short, "if an infinite regress is to be avoided, there must be a first cause of movement that causes movement not by virtue of being moved by anything else, in other words, which is either self-moved or unmoved."[11]

Cosmic Orchestration: Growing Complexity

We are able now to grasp the layout of Aristotle's cosmos in the *Physics*. Aristotle conceives of the universe as an immense sphere within which, at regular distances, ever smaller rotating hoops are located and which all adopt the circular rotation of the outer sphere. This results in a cluster of concentric glassy shells (technically: hollow globes) with each one inside another except for the outermost, and arranged in such a way that the smaller they become, the closer also they enclose the Earth, which is in the middle of the universe. Closest to the Earth are the three sublunary circles, which do not rotate and which contain the four elements. Water (liquid) is constitutive of the first circle, air (gaseous) of the second, and fire (radiant) of the third. Above the three sublunary circles, one has the seven planetary hoops, each with their own planet. First comes the orbit of the Moon (fourth circle), then that of the Sun (fifth circle), Mercury (sixth circle), Venus (seventh circle), Mars (eight circle), Jupiter (ninth circle), and finally that of Saturn (tenth circle). All these circles are crowned by the firmament or the *primum mobile*, which carries the fixed stars. This vault rotates in twenty-four hours around the Earth and imposes its westbound movement on the rotations of the planetary circles.

---

11   William Ross, *Aristotle's Physics: A Revised Text with Introduction and Commentary* (Oxford: Clarendon Press, 1936), 89.

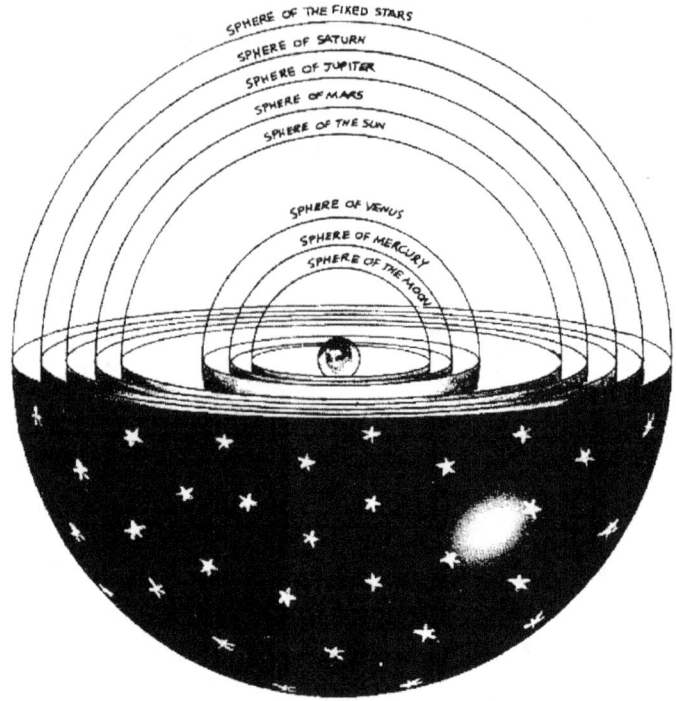

Figure 4. Planetary spheres
Source: Stephen W. Hawking, *A Bief History of Time* (London: Bantam, 1988), 3.
This figure is reproduced courtesy of Professor Stephen W. Hawking,
University of Oxford.

For Aristotle it was evident that the planetary circles would all naturally take on this westbound rotation of the firmament, since this was in line
with his theory of the transmission of kinetic force through close physical
contact. With the firmament moving westbound, the direction of this movement is directly passed on to the circle of Saturn, which passes it on to the
circle of Jupiter, which passes it on to the circle of Mars etc....till the whole
clockwork of celestial rotating circles takes on the westbound direction. Yet,
after a while Aristotle began to realize that this schema was too simplistic,
since astronomical data showed that the seven planets were at the same time
carried along the ecliptic in a reverse movement (from the west to the east),
which made them progressively lag behind the firmament's westbound rotation. Moreover, the planets' route along the ecliptic formed an angle of 23.5°
with the equator of the universe, elements Plato had already brought up in
the *Timaeus*. So, the system of concentric glassy shells had to be refined. This
is what Aristotle undertook in his work *Metaphysics*. There he, first of all,

gave a detailed account of the system of planetary circles or spheres developed by Plato's disciple Eudoxus, followed by his own suggestion of how to improve it. I begin with Aristotle's account of Eudoxus' system.

Eudoxus of Cnidos (408–355) was the first to offer a model whereby the apparently complex motions of the heavenly bodies were seen as resulting from a combination of regular circular motions. He broke up the apparently capricious motions into their constituent parts. This led to a multiplication of the spheres. Instead of allotting one ring to each planet, he worked out a system in which each planet was attached to the innermost sphere of a cluster of nested or homocentric spheres (the one being inside the other). In this way the composite motion of the particular celestial body was dissolved into a combination of regularly rotating circles, each with their own direction, speed, and axial inclination. The aim and purpose of this grouping was to give an explanation for observed facts.

Eudoxus needed three nested or homocentric rotating spheres to explain the particular behavior of the Moon and the Sun, whereas for the remaining planets—which unlike Moon and Sun exhibited particular loopings—he needed four nested or homocentric rotating spheres (see more details below).

(i)All in all Moon, Sun, and planets have the following features in common: (a) their outermost nested sphere adopts the westbound movement of the firmament, whereas (b) their second nested sphere takes on an eastward rotation along the path of the ecliptic.[12] The outermost nested sphere thus wholly espouses the firmament's westbound rotation, which accounts for the succession of days and nights, whereas the second nested sphere takes on that movement and adds to it its own input: the eastward oblique journey of the planets along the ecliptic (it is on this trajectory that the Sun marks the succession of seasons). As already mentioned in the analysis of *Timaeus* it takes the Moon only a month to traverse its path along the ecliptic, whereas for the Sun this travel takes 365 days. For Mercury and Venus this revolution is shorter than a year, but for the "outer planets" it takes years to traverse the path along the ecliptic: 2 years for Mars, 12 years for Jupiter, and 30 years for Saturn.

(ii) Besides these common features there are still further particularities to be explained. In order to account for the particular behavior of the Moon and the Sun, a third nested sphere will be needed. In the course of its monthly

---

12   In Aristotle's words: "The outermost sphere adopts the movement of the fixed stars," whereas the second sphere inside the first "revolves in the circle which bisects the zodiac" (*Metaphysics* XII, viii, 1073b 19–20).

journey along the ecliptic, the Moon crosses the equator of the universe at two points, called nodes: the ascending node when the Moon passes from south to north and the descending node when it passes from north to south. These nodes slowly displace themselves backward; it takes 18 2/3 years before the original position of the nodes is reached again.[13] In order to account for this recession of the nodal points, Eudoxus assumed a third nested sphere (inside the second) which modifies the oblique eastward motion of the second nested sphere. A similar explanation is needed for the Sun. About the 2nd of January the Sun appears to increase in velocity, and to decrease in velocity six months later, a phenomenon which refers to the lengthening and the shortening of the days. In order to account for this phenomenon, Eudoxus also assumed a third nested sphere. In both cases the third nested sphere rotates, like the second one, from the west to the east but with a different axial position when compared to the former.[14]

The remaining planets show, besides their combined revolutions from the east to the west (the direction of the firmament) and from the west to the east (along the ecliptic), the following common characteristic:

> [They] seem to slow down at irregular intervals in their courses, remain stationary for varying periods [...], and then move westward (retrograde in the zodiac) for a time, stop again, and finally resume their eastward motion, at the same time exhibiting changes of latitude—thus in effect executing *elongated loops* in the sky relative to the fixed stars.[15]

To explain the phenomenon of looping, two supplementary nested spheres are needed, one to interrupt the steady rhythm of the oblique eastward movement and one to obstruct the interruption.

This gives the following schema of nested spheres for Mercury, Venus, Mars, Jupiter, and Saturn. The first (outermost) nested sphere espouses the westbound daily revolution of the firmament. The second nested sphere accounts for the planet's journey along the ecliptic. The combined motion of the third and fourth nested sphere explains the looping of the planet. Eudoxus reached this effect by giving these spheres a different axial

13   Jacques Merleau-Ponty and Bruno Morando, *Les trois étappes de la cosmologie*, 64–65.

14   In Aristotle's words: "the third [sphere of the Sun] revolves in a circle which is inclined across the breadth of the zodiac; but the circle in which the Moon moves is inclined at a greater angle than that in which the Sun moves" (*Metaphysics*, XII, viii, 1073b 21–23).

15   D. R. Dicks, *Early Greek Astronomy to Aristotle* (Bristol: Thames and Hudson, 1970), 25.

rotation, and making them rotate in opposite direction but with equal speed.[16] Computer steered animations show that the interaction between them results in a wobbling figure-eight shape, technically termed hippopede: looked at from the Earth the particular planet, say Mars, which is located on the inner nested sphere, appears to continuously wobble back and forth on its path along the ecliptic.

Eudoxus' construction results in a sum total of 26 spheres. (4 + 4 + 4 + 4 + 4 + 3 + 3 = 26). Later on Callippus (born *c.* 337 BC) increased Eudoxus' number of spheres by adding two more spheres to the Sun and the Moon, and one more sphere to Mars, Venus, and Mercury. He did this with a view of getting a still better comprehension of the specific behavior of each planet. This way, the sum total of spheres added up to 33, a number that Aristotle is going to expand to 55.

	Eudoxus	Callippus	Aristotle
Saturn	4 spheres		+ 3 spheres
Jupiter	4 spheres		+ 3 spheres
Mars	4 spheres	+ 1 sphere	+ 4 spheres
Venus	4 spheres	+ 1 sphere	+ 4 spheres
Mercury	4 spheres	+ 1 sphere	+ 4 spheres
Sun	3 spheres	+ 2 spheres	+ 4 spheres
Moon	3 spheres	+ 2 spheres	
	————————	————————	————————
	26 spheres	+ 7 spheres	+ 22 spheres = 55 spheres

Aristotle welcomed the fact that Eudoxus and Callippus, with the help of clusters of often counter-rotating spheres, succeeded in fairly explaining the composite motion of each of the planets. Yet, he found that this procedure raised a serious problem. Given the close physical contact of all the clusters of spheres, one could expect Saturn, the outermost planet, to impose its rhythm of looping on all the inner planets. Likewise, one could expect Mercury to pass on its looping to the Sun, which manifestly does not occur. To resolve this difficulty Aristotle "considered it necessary to prevent the motion of an outer group of nested spheres from being transmitted to the inner groups; this was attained by inserting a number of neutralizing spheres, the number of which is each time one less than that of the preceding group (for the daily

---

16   See Aristotle, *Metaphysics*, XII, viii, 1073b 29–31: "The third sphere of all the planets has its poles in the circle which bisects the zodiac; and the fourth sphere moves in the circle inclined to the equator of the third."

rotation may indeed be transferred inward)."[17] In this way factual buffer zones were created to shield each planet from automatically adopting the rhythm of its higher neighbor. Let us note that these neutralizing spheres do not properly produce any motion; they rather cancel an existing particular motion by rotating in the opposite direction. In this way each cluster of nested spheres was given a chance to start from scratch, so to speak, not hindered by the particular behavior of a more highly located cluster of nested spheres.

The clusters of nested spheres that carry Saturn and Jupiter are each given 3 neutralizing spheres, whereas four neutralizing spheres are allotted to the clusters of nested spheres that carry the 4 next planets: this gives a total number of 22 neutralizing spheres (6 + 16 = 22). They are placed underneath the nested sphere that immediately carries the planet in question. With reference to Callippus' system, this means that the neutralizing spheres are right under the fourth nested sphere of Saturn and Jupiter, and right under the fifth nested sphere of each of the other planets. It is easy to calculate the number of neutralizing spheres: each time one less that the number of nested spheres in a planetary cluster. The outermost sphere of a cluster of nested spheres ought not to be cancelled, since it espouses each time the westbound rhythm of the firmament. Likewise, the Moon, the lowest planet, does not require any neutralizing sphere since it is at the bottom of the descending hierarchy of celestial bodies.

The Concept of God in *Metaphysics*. The Highest God and His Auxiliary Movers: The 55 Cosmic Intelligences

Aristotle's refinement of Eudoxus' and Callippus' system of the planetary spheres is to be found in his *Metaphysics*. In this work, too, he begins to realize that the Prime Unmoved Mover, who sets the firmament in motion, must be of a totally different nature than the things he moves. Up to now Aristotle took it for granted that the Prime Unmoved Mover gave a kinetic thrust to the firmament by using, so to speak, his hands to make the firmament rotate. Transmission of motion through close physical contact was the rule of thumb in his work *Physics*. In his *Metaphysics*, however, Aristotle brings to the fore a totally immaterial God, who moves the firmament by exerting a seductive lure on it: "God causes motion as being an object of love (*kinei hoos eroomenon*)" (*Metaph.* XII, vii, 1072 b 4–5).[18] Moreover, there is a

---

17   E. J. Dijksterhuis, *The Mechanization of the World Picture*, trans. C. Dikshoorn (New York: Oxford University Press, 1969), 34.

18   Aristotle, *Metaphysics*, ed. Hugh Tredennick, The Loeb Classical Library (London: Heinemann, 1969).

second indication that Aristotle abandoned the rule upheld in the *Physics* that motion is transmitted through close physical contact. In his *Metaphysics* he admits of the existence of 55 unmoved movers, besides the Prime Unmoved Mover, their number corresponding to the 55 spheres that are required to explain the particular behavior of the planets. They all move their respective sphere, not through physical contact, but through the energy and seductive power they possess as immaterial entities.

A striking feature is the immateriality ascribed to both the prime mover and the 55 auxiliary movers. The first time Aristotle brings up the concept of an immaterial Prime Mover is in his work *On the Heavens*. There he points out that the firmament, set in motion by the Prime Mover, is "ungenerated, indestructible, and altogether changeless," since it is made of quintessence. Yet, he adds, because of its rotation the firmament is caught up in spatial motion. In contrast to it, the prime mover is not characterized by spatial motion: he is totally immaterial. That is why he "has far better reason to be ungenerated, indestructible, and altogether changeless" (*De Coelo*, II, vi, 288b 1–8).

In his *Metaphysics* Aristotle upholds this insight, but goes a step further: he sets out to multiply the number of cosmic movers by positing 55 incorporeal movers, each of them allocated to one of the 55 planetary spheres. He writes:

> Since we can see that besides the simple spatial motion of the universe (which we hold to be excited by the primary immovable substance) there are other spatial movements—those of the planetary spheres—which are eternal (because a body which moves in a circle is eternal and is never at rest), then each of these spatial motions (*phorai*) must also be excited by an [incorporeal] substance which is essentially immovable and eternal.

These incorporeal substances are as eternal as the planetary spheres, but unlike these spheres

> they are essentially immovable, and without magnitude [...]. Thus it is clear that the movers are [incorporeal] substances and that one of them is first and another second and so on in the same order as the spatial motions of the heavenly bodies. (*Metaph.* XII viii, 1073a 30–1073b 3)

Their number corresponds to the number of planetary spheres: "This, then, may be taken to be the number of the spheres [55 on the whole] and thus it is reasonable to suppose that there are as many [immaterial] immovable substances and principles" (*Metaph.* XII viii, 1074 a 13–17).

One might ask what exactly motivated Aristotle to allot an incorporeal cosmic mover to each of the 55 planetary spheres? The basic reason, as already mentioned, is that he abandoned the principle of the transmission of motion through close physical contact. The diverse motions of the 55 planetary spheres could no longer be explained by the Prime Mover's sole kinetic thrust forced upon the firmament, from where it rolled down to set the other spheres in motion. If one takes seriously the whole array of rotating and counter-rotating planetary spheres, then the hypothesis of a *single mover* that excites the rotation of the highest firmament and through it also that of all the lower orbiting spheres, must of necessity collapse. In order to honor the irreducible character of each sphere—of those within the clusters of nested spheres as well as of the neutralizing spheres—Aristotle had thus to posit the existence of a proper immaterial mover allotted to each of them.

Just like the highest God, the Prime Mover, the 55 immaterial movers all lack spatial extension; they are all without magnitude. In antiquity it was believed that the more immaterial an entity was, the more reality it possessed. Originally, this supra-spatial domain was not part of Aristotle's thought, but in the *Metaphysics*, he ventured to explore it, for the simple reason that on the basis of a composite system of corporeal spatial rotations no ultimate coherence could be given to the universe. Only the concerted activity of immaterial movers, all of them having a specific effect on the rotation of their particular sphere, was to assure this ultimate coherence.

The 55 immaterial movers, also termed "immaterial substances" or "cosmic intelligences," ought to be imagined as intensive energetic points without spatial extension. As condensed nuclei, they draw their life force from within. They are—at their level of influence—pure "act" (*actus*): not standing in need of having to pass from potentiality to actuality. From eternity they enjoy the full possession of their own substantial being. Unlike the rotating spheres, whose activity is broken up in spatial sequences, they are endowed with a compact energy. This allows them to animate the rotation of their planetary sphere without losing the least fraction of their full energy. If such gradual loss were to take place, the perpetuity of the celestial rotations would be at stake. This explains why Aristotle sets great store by the immaterial substances: their realm wholly transcends the quintessential domain of the celestial spheres and bodies, whose eternity, so to speak, depends on theirs.

Aristotle is rather silent about how this immaterial realm is internally structured. He only suggests that a harmonious bond exists between the immaterial substances, without specifying how the 55 cosmic intelligences relate to the Highest Cosmic Intelligence. This is a question Thomas Aquinas (1227–1274) will tackle when resolving the disputed question, in the Middle Ages, about the "eternity of the cosmos" (see below). In the meantime, we get a precise picture of the self-awareness of the Highest Cosmic Intelligence above the firmament, the prototype after which the 55 cosmic intelligences are modeled. In *Metaphysics* XII, vii Aristotle portrays how the highest God, from eternity to eternity, beholds the whole range of his thought (*noesis noeseoos*) and immensely enjoys this spectacle.

Aristotle's famous definition of God reads as follows:

> [In God] thought and the object of thought are the same (*tauton nous kai noeton*) […], for thought actually functions when it possesses this object. Hence it is actuality rather than potentiality that is held to be the divine possession of thought, and its active contemplation is that which is most pleasant and best. […] If, then, the happiness (*eudaimonia*) which God always enjoys is as great as that which we enjoy sometimes, it is marvelous. And if it is greater, this is still more marvelous. Nevertheless it is so. Moreover, life belongs to God. For the actuality of thought is life, and God is that actuality; and the essential actuality of God is life most good and eternal. We hold, then, that God (*ho theos*) is a living being, eternal, most good; and therefore life and a continuous eternal existence belong to God; for that is what God is. (*Metaph.* XII, vii, 1072b 22–31)

It goes without saying that this definition of God will be welcomed in the Christian milieu, especially by Thomas Aquinas.

In a next step Aristotle sets out to explain how the happiness experienced by God arouses in his surroundings a desire for sharing in this bliss. The divinity's happiness lures the ranks underneath It into desiring a comparable happiness and perfection. They seek, in other words, to imitate the deity (*homoioosis theooi)* and to acquire—on their respective echelon of being—an excellence that is analogous to that of the deity. In technical terms: God is the "final cause" attracting the entities in the universe to their utmost perfection. The starry firmament is eager to carry out its regular rotation as elegantly as possible, as it is lured to this excellence by the Highest Cosmic Intelligence, whose stupendous happiness exerts an irresistible attraction on all the entities around it. "God causes motion as an object of love (*kinei hoos eroomenon)*" (*Metaph.* XII, vii, 1072 b 4–5).

Attraction by divine lure is something totally new in Aristotle's theory of motion, especially when compared with the kinetic approach of the *Physics*, where things were set in motion because of the motive thrust that God, the efficient cause, gave to the firmament, which in turn passed on this kinetic force to the lower spheres.This mechanical transmission of force is replaced now with a creative awakening of motion on the basis of enchantment and seduction. The Highest Unmoved Mover excites the rotation of the firmament

> not by doing something Himself; that would be contrary to the fact that He is *actus purus* [pure actuality], and thus has nothing left to realize; perfect actuality must consist in absolute inactivity. However, He moves the firmament, as the Aristotelian formula says *hoos eroomenon*, by being loved, i.e., the motion is the consequence of the affection in which He is held by the firmament, of the craving for perfection that He arouses in it.[19]

From this model, one may deduce that the 55 cosmic intelligences do exactly the same. They, too, exert a lure on their respective sphere in the various clusters of spheres around the planets, with the effect that all these spheres begin to be seduced into carrying out their perfect circular rotations with delight. It is as if a golden chain of seductive energy is descending continually from the Prime Cosmic Intelligence to the 55 cosmic intelligences—from the highest to the lowest—until it reaches the lowest sphere to which the Moon is attached. There the golden chain stops and the domain of uncertainty and perishing begins.

Aristotle's concept of God in the *Metaphysics*—of the final cause that acts upon the cosmos through seduction—crowns his meditative cosmology. In this form it will be handed on to medieval theology: "God is the Love who moves the Sun and the stars," writes Dante. The Prime Unmoved Mover exerts such a seductive power that It lures the whole firmament with the fixed stars towards performing its perfect circular movement. God's seductive power, then, flows further down under the firmament, where It invites the other unmoved movers (cosmic intelligences) to exert their seductive power on their respective spheres around the planets. In technical terms, the Prime Mover encourages the other unmoved movers to make use of their seductive power as "secondary causes" in the cosmos. Through their additional cosmic lure, they begin to unpack, so to speak, the cosmic lure of the Prime Unmoved Mover. This diversification results in the miracle of an array of rotating spheres that gives birth to the perennial splendor of a varied yet harmonious heavenly ballet. In the sublunary domain begins the capricious realm of the things that are perishable. It is separated by a gulf from the heavenly ballet, which nonetheless, in a mysterious way, secures the steady reproduction of life on Earth.

---

19   E. J. Dijksterhuis, *The Mechanization of the World Picture*, 35 .

CHAPTER 4

Christianity's Assimilation of Greek Cosmology

Patristic Period

In the proclamation of the gospel to the Greek-speaking Gentiles two methodological principles were utilized that directly related to cosmology. First, it was recognized that the natural world exhibits traces of the Creator (*vestigia Dei*). The fact that God exists can be derived from the study of nature: "Since the beginning of the world, God's invisible qualities—his eternal power and divine nature—have been clearly known, being understood from what has been made" (Rom 1:19). Second, to counter the pessimistic view of Gnosticism with its stress on sinful character of material existence, the Church Fathers affirmed the goodness of the created realm.

This positive appreciation of the created realm allows us to understand why the ancient church initially did by no means discourage the study of the basic structures of nature, in as much as it prepared one to embrace the gospel values with deeper conviction (*praeparatio evangelii*). The church leaders perfectly understood the extent to which the faithful may strengthen their faith in God, the creator and redeemer, through an intense study of the Greek classics of the philosophy of nature. In the Patristic period, this policy led to the elaboration of the two books of theology: in order to understand the fullness of God's self-revelation one should read together the "book of revelation" and the "book of nature." In the words of St. Augustine: "It is the divine page that you must listen to; it is the book of the universe that you must observe. The pages of Scripture can only be read by those who know how to read and write, while everyone, even the illiterate, can read the book of the universe" (*Enarrationes in Psalmos,* 45,7).

As a matter of fact the Church Fathers could rely on a tradition that was already initiated by the Jewish philosopher Philo of Alexandria (20 BCE–50

CE), who saw in the term "logos" the confluence of Jewish and Greek thought. There was, indeed, the happy coincidence that the term "logos"—which in the Jewish wisdom literature referred to the creative Word of God, and which in the Greek stoic milieu stood for the all-organizing immanent principle in the world—could be used as a linguistic vehicle allowing one to travel to and fro between the two domains of cosmic piety and the belief in the biblical God of salvation. Christian theologians, like Justin (100–165), Clement of Alexandria (150–215) and the author of the letter to Diognetus (2nd–3rd cent.) had no problem at all with juxtaposing passages from the creation story in Genesis and from Plato's Timaeus: admiring in both sources the might of the Creator who brought about order out of chaos.

An uncompromising repudiation of Greek cosmology only arose from the moment when Christian apologists realized that on account of their study of the basic structures of the universe a many great Hellenistic intellectuals refused to embrace Christian faith. This encounter with 'unbelief' was a motive for them to critically examine the cosmological presuppositions of the Gentiles. In spite of their deep admiration of Plato, they also began to question him for admitting of the eternal existence of matter in his creation story, for the doctrine of the co-eternity of matter with God served to undermine belief in the Creator's omnipotence. One of the most eloquent texts is the refutation of Plato by Theophilus of Antioch in the 2nd century:

> Plato and his disciples acknowledge an uncreated God, father and maker of the universe; at the same time they maintain that matter, just like God, has no origin, and is as eternal as God. But if God has no beginning and matter has also no origin, then God cannot possibly be the [sovereign] maker of the universe. So God's absolute power is at stake [...] In reality, God's omnipotence manifests itself in all the things He made out of nothing. (*Ad Autolicum* II, 4; PG. 6; col. 1051–1052)[1]

As a matter of fact, the apologists had no problem at all with the platonic ideas, which in Middle Platonism—the interpretation of Platonism that reigned during the period in which they lived—were regarded as being located within God's mind.

Aristotle's cosmology, however, also appeared, on closer inspection, to be defective, for to the extent that his theory suggests that next to the highest God, who lures the firmament into its perfect rotation, there are 55 auxiliary cosmic intelligence that are doing the same with respect to their

---

1   Quoted in Paul De Haes, *Schepping en schepsel* (Tielt & Den Haag: Lannoo,1966), 23
     (Translation mine).

planetary sphere, this view threatens to reduce God to a world-immanent principle of motion. Moreover, the Aristotelian God turns out to be lacking in the deep emotions that are indicative of a genuine personality. Although, as an object of love, he exerts a lure on the firmament, God is not deeply concerned about the plight of humankind in history.

A later example of a potent voice of protest against Greek cosmology is St. Ambrose (339–397). In the first chapter of the *Hexaemeron* he writes:

> To such an extent have men's opinions varied that some, like Plato and his pupils, have established three principles for all things; that is, God, Idea and Matter. The same philosophers hold that these principles are uncreated, incorruptible, and without a beginning. They maintain that God, acting not as a creator of matter but as a craftsman who reproduced a model, that is, an Idea, made the world out of matter. This matter, which they call *hyle*, is considered to have given the power of creation to all things. The world, too, they regard as incorruptible, not created or made. Still others hold opinions such as those which Aristotle considered worthy of being discussed with his pupils. These postulate two principles, matter and form, and along with these a third principle which is called "efficient," which Aristotle considered to be sufficient to bring effectively into existence what in his opinion should be initiated. What, therefore, is more absurd than to link the eternity of the work of creation with the eternity of God the omnipotent? Or to identify the creation itself with God, so as to confer divine honors on the sky, the earth, and the sea? [...] How is it possible to arrive at an estimate of the truth amid such warring opinions? Some, indeed, state that the world itself is God, inasmuch as they consider that a divine mind seems to be within it, while others maintain that God is in parts of the world. [...] Under the inspiration of the Holy Spirit, Moses, a holy man, foresaw that these errors would appear among men and perhaps had already appeared. At the opening of his work he speaks thus: "In the beginning God created heaven and earth." He linked together the beginning of things, the Creator of the world, and the creation of matter in order that you might understand that God existed before the beginning of the world or that He was Himself the beginning of all things. (*In Hexaemeron* I, 1–4)[2]

One will have to wait till the widespread reception of the new synthesis of Plato and Aristotle developed by Plotinus (205–270) before Christian theology will altogether begin to inculturate itself into the intellectual climate of Greek cosmology. Plotinus, who lived in Alexandria but later founded a philosophical school in Rome, succeeded in elaborating a special notion of

---

2   Ambrose, *Hexaemeron, Paradise, and Cain and Abel*, trans. John Savage (New York: Catholic University of America, 1961).

causation—coupled with emanation—that safeguards the transcendence of the first cause (the One) over all the entities that participate in the outflow of Its being. This view had tremendous implications both for the domain of spirituality and of cosmology. On the level of spirituality, the One was experienced as the ineffable presence that accompanied all human striving, whereas, on the level of cosmology, the One was honored as the ungraspable fountainhead from which the formative principles of the cosmos emanate. In both cases, the reality of the One could only be evoked through thoughtful negations: It (the One) is not "being" or "life" in the common understanding of these terms, since It is basically "beyond" all this. This explains, for example, the cascade of negations in *The Mystical Theology* of Ps-Dionysius the Areopagite, a Christian monk of the 5th or 6th century, who assimilated Plotinus through Proclus (412–485), the system builder of Neoplatonism, as the movement was soon to be called:

> Neither is God darkness nor light, nor the false nor the true; nor can any affirmation or negation be applied to him, for although we may affirm or deny the things below Him, we can neither affirm nor deny Him, inasmuch as the all-perfect and unique Cause of things transcends all affirmation, and the simple pre-eminence of His absolute nature is outside of every negation—free from every limitation and beyond them all.[3]

In Plotinus' *Enneads* the One (*to Hen*) is presented, on the cosmological level, as the unique Cause that, through a hierarchical order of emanations, generates all things, and in doing so remains elevated above these emanations, be it the cosmic intellect (*Nous,* the first emanation) or the World Soul (*Psyche,* the second emanation):

> The One is not being but the generator of being. This is the first act of generation, as it were: being perfect because It seeks nothing, has nothing, needs nothing, the One, so to speak, overflows and Its overfullness had made another; and what has come to be turned back to It [i.e. to the One] this came to be intellect'(*Nous*). (*Enneads* V, 2 (11), 1, 5)[4]

---

3   Ps-Dionysius Areopagite, *Mystical Theology*, in *Mysticism: A Study and an Anthology*, F. C. Happold (Harmondsworth: Penguin Books, 1975), 217.

4   See Kevin Corrigan, *Reading Plotinus: A Practical Introduction to Neoplatonism* (West Lafayette: Purdue University Press, 2005), 13.

In this succinct text various elements flow together: (i) The One is beyond any multiplicity, whereas that which It generates will develop an activity within the multiple. (ii) The One does not stand in need of anything; yet, out of pure generosity and overfullness It gives birth to other things. (iii) The One is always perfect and therefore produces everlastingly; and Its product is less than Itself. Nothing can come from the One except that which is next greater after It. *Nous* is next to it in greatness and second to It. In a sense the One never gives to the lower ranks what It properly possesses: Its utmost simplicity and non-dividedness.

Decisive is the way in which Plotinus describes the emanation of the intellect (*Nous*): "what has come to be turned back to It [i.e., to the One] this came to be intellect." In a prima facie consideration, it might seem as if the intellect got seduced by multiplicity, and lapsed into discursive thinking because it fell in love with a variety of objects that challenged its curiosity. Yet, this is not exactly what happened. Plotinus makes it clear that when the One generates "another" (*Nous*), this "other" turns, out of sheer delight, its gaze back to the One; but since the One now appears to the intellect as an "object of knowledge" it can only approach it discursively, without being able to grasp it in its "simple being": "From the perspective of the intellect which looks towards the One, the One becomes an intelligible object (*noeton*). Yet the One as perceived is not identical with the One as It is [...] the intellect cannot apprehend the One as it is in Itself, i.e., as absolutely simple."[5]

Also the World Soul (*Psyche*) is an emanation of the One, although its emergence occurs through the mediation of *Nous.* The World Soul will have to animate the body of the cosmos; it becomes operative in that body because it is empowered to it by *Nous.* In this activity Soul feels the stimulus of *Nous,* while being unable to experience what *Nous* is in itself. Insofar as it is turned towards *Nous,* Soul perceives a glimpse of *Nous*—never the latter's discursive operation; in as far as it is turned away from it, it departs from the intelligible world and is, with "its lower part," engaged in the world of bodies. Let us recall that, in the *Timaeus,* the Demiurge shaped the World Soul as a firmly knotted nervous system, endowed with a quasi-indestructible energy for immanently acting upon the material entities. The World Soul, in turn, gives birth to the separate substances and finally to matter that is going to receive the imprint of the eternal ideas contained in the Nous.

Plotinus's system of emanations would be misunderstood if one were to focus only on the "diminishment by degrees"—in the successive lower

---

5   David Rehm, *Plotinus' Use of Dunamis and Energeia in his Account of Emanation from the One* (Chicago: The University of Chicago Press, UMI Dissertations, 1994), 46.

levels—of the unified energy of the One. Such a "diminishment by degrees" is, to be sure, taking place, but is opulently recompensed by the unbroken empowerment that overflows from the One. *Nous* and *Psyche* are, basically, active principles sharing in the One's energy—which they activate at their own level. Their awareness of participating in this creative power generates in the emanations (hypostases) a feeling of assurance, which is intensified the more they celebrate their "return" to the overwhelming Origin. To the emanating "outflow" (*exitus*) naturally corresponds a movement of ecstatic "return" (*reditus*).

It is this mystical background that appealed to a great many Christian theologians: In the Greek-speaking world the Enneads were read by the theologians Eusebius of Caesarea, Basil the Great, and his younger brother Gregory of Nyssa, and their friend Gregory Nazianzen in the fourth century. St. Augustine (354–430) read Plotinus in the Latin translation of Marius Victorinus; the reading had such an effect on him that he abandoned Manichaeism, the Gnostic school to which he had been an adherent for nine years.

The Nicene Creed

Much earlier than Plotinus, emanations were already in vogue in the Gnostic mythological systems of Basilides (ca 117–138) and Valentinus (ca 100–ca 160). Yet, Gnostic emanations, instead of proceeding from the generosity of the One, as in Plotinus, were linked up to a moment of fall. It is through the sinfulness of Sophia, one of the lowest emanations, that the world of the humans came into existence with its exposure to sensuous temptations. So, salvation will consist in freeing oneself from the despised body. Also in the Christian milieu emanations had been used, mainly to explain the relations, in the triune God, between Father, Son, and Holy Spirit. A fine example of this type of theologizing is Origen (185–254). For him, the Father is, just as later in Plotinus, conceived of in terms of utmost simplicity. He is the One (*to Hen*) "who in a timeless act of creation [i.e., emanation] begets the second God, the wisdom of God, also termed logos or word [...] In the logos all the ideas or the intelligible world are present."[6] The Son is, thus, associated with the realm of ideas that Plato identified in the *Timaeus*; he is also given some of the characteristics of the World Soul, insofar as he animates the beings endowed with intellect and soul to discover their spiritual roots in the One.

---

6   H.J Kraemer, *Der Ursprung der Geistmetaphysik* (Amsterdam: Gruener Publisher, 1967), 286–87 (translation mine).

Origen was twenty years older than Plotinus; some scholars opine that he must have had the same teacher as Plotinus, Ammonius Saccas. Although some similarities between Origen and Plotinus can be established, it is clear that Origen deviates from Plotinus in that—in line with the Jewish-Christian tradition—he maintains that the material world has been created out of nothing. On the other hand, Origen will be critiqued for subordinating the Son to the Father, something that is self-evident from the Plotinian perspective: namely that "nothing can come from the One except that which is next greater after It."

In light of these considerations it is worthwhile to have a look at the Nicene Creed: To what extent does it make use of emanations? How does it exactly explain the creation of the world? When I say the "Nicene Creed" this ought to be specified. The "Nicene Creed" we recite in Catholic and Byzantine liturgy is not the rather short text promulgated in the council of Nicea (325), but the longer version that was agreed upon in the council of Nicea-Constantinople (381) and which also includes the formulation of the procession of the Holy Spirit (a formulation that will be slightly changed in the Latin church). It is this longer text that I will use in my comments on the creed.

In the Nicene Creed, the process of emanation is restricted to the birth of the Son from the Father, and the issuing forth of the Spirit from the Father through the Son[7] (in the version of the Latin Church: from the Father and the Son). Only then, after this twofold emanation, the "Father, the Almighty" will create the world—out of nothing—through the mediation of the Son. As to the emanation process, ample space is given to the generation of the Son:[8] "the only Son of God is eternally begotten of the Father" and "is of one being with the Father" *(homoousion tooi patri)*. To avoid the impression that Christ might be regarded as a common creature, the creed affirms that the Son "is begotten, not made" *(gennethenta ou poiethenta)*. That the Son "is not made" means that he has not been brought forth in the mode in which finite entities are produced "out of nothing." In contrast to the latter, the consubstantial Son is entirely located at the side of the Father, the *pantokratoor.* He is "God from God" *(theon ek theou)* "light from light" *(phoos ek phootos)*, "true God from true God" *(alethinon theon ek alethinou theou)*. This imagery clearly points to a process of emanation. That we have to do with an emanation in the style

---

7   I follow here the later Byzantine reading of procession of the Spirit.

8   The reason for this is the Arian controversy. Arius had maintained that "before he was begotten the Son was not" and "that he came to be from another substance than the substance of the Father."

of Plotinus is further confirmed in the Greek formulation of the procession of the Holy Spirit: "We believe in the Holy Spirit [...] who proceeds from the Father *through* the Son" (*dia tou huiou*). Indeed, if the energy of the Father is to reach the Holy Spirit it must pass through the mediation of the Son. In the logic of Plotinus the emanations take place in the mode of a cascade, one hypostasis flowing from the other. Let us note in passing that the Greek theologians were flabbergasted at the alternative reading of the Latin Church, namely that the Holy Spirit proceeds from the Father *and* the Son.

When it comes to the creation of the world—out of nothing—the *dia*-formula appears again: "through him [the Son] all things were made" (*di'hou ta panta egeneto*). Here, too, the mediating role of the Son is highlighted: The Father is the maker of all things, visible and invisible, but the Son is the one who mediates this creative activity and renders it specific. Because this is such a crucial notion, I would like to problematize it from our cultural situation of today. Taken in itself, the expression "through him all things were made" can be read in a twofold way. It can be read as referring to the order of salvation: no saving event can take place except through the mediation of Christ, who is totally at the side of God, and who, in fact, wrought our salvation through his incarnation in the human flesh (the creed dwells on the importance of Christ's death and resurrection). Yet, "through him all things were made" can and must also be read as referring to the creation of the universe. It is at this level that contemporary people will have problems; they will find it difficult to visualize in what concrete fashion the Son has been instrumental in effecting the material production of the universe. When applied to the physical creation of the world, the dictum *di'hou ta panta egeneto* sounds rather enigmatic, not in the least for whoever is somewhat acquainted with the astrophysical origin of the evolutionary processes that have led to the emergence of beings like us. How to visualize the Son's (the Logos') mediation in steering the Big Bang and the astral formations that came out of it?

Given the scarcity of information about the Son's mediation in the act of creation, some theologians—Pierre Teilhard de Chardin is one of them—set out to place Christ's role in the formation of the universe in an evolutionary perspective. For Teilhard, "the great cosmic attributes of Christ, those which (particularly in St. John and St. Paul) accord him a universal and final primacy over creation, these attributes . . . only assume their full dimension in the setting of an evolution that is both spiritual and convergent."[9] Based on his studies of paleontology, Teilhard took it for granted that matter is evolving

---

9   Pierre Teilhard de Chardin, in *Catholicism and Science*, 1946, IX, 189.

towards consciousness and self-consciousness, and that this evolution will reach its apogee in the Omega point, where all the entities in the universe will be united in mutual love—the indication of Christ's final primacy over creation. The Omega point is the culmination of a long evolution that started with the emergence of pre-life (Alpha), in which bio-molecules were formed, to then reach the stage of life (from viruses to animals with skeletons, and then mammals) and that of self-consciousness (from pre-hominians to homo sapiens) to eventually attain the stage of human planetization (with love-energy as unifying force). The Omega point of convergence is nothing else but the cosmic Christ who exerts a potent lure on the whole of the evolutionary process, so that it may attain its full completion. In this schema one perceives a shift in emphasis from "origin" to "finality," so that the burning question now becomes: "*in view of what*," "*for which motive*," and "*for what purpose*" has God ventured to call the universe into existence?

In this light one may ask the question whether Christians today would not find a stronger inspiration in the Nicene Creed, if "through him" (*di' autou*) would be replaced with "in view of him" (*eis hon*), in accordance with the Letter to the Colossians in which it is stated: "the whole universe has been created through him (*di'autou*) and in view of him" (*eis auton*). The confession of faith would then read as follows: "We believe in one Lord Jesus Christ, the only Son of God, eternally begotten of the Father, God from God, light from light....In view of whom all things were made." Christ, the logos, is seen as installing in the natural world a dynamism of growth and development that has the tendency to encompass the whole universe. Not only did the logos take on human flesh, he also assumed a cosmic dimension in that he lures the whole of the evolving world towards its full completion.

Teilhard's vision of Christ as the Omega point was the result of his field work as a paleontologist, which sensitized him to the evolutionary processes in nature. So, it would be totally unrealistic to expect the bishops who participated in the deliberations of the council of Nicea to espouse such an evolutionary vision. On the other hand, one can presume that a great many of them were familiar with Greek cosmology and the role it attributed to the logos or the *Nous* in structuring the world. According to the Stoa, logos is a world-immanent cosmic principle, whereas *Nous,* in Plotinus, is the seat of the mathematical proportions on which the finite entities are modeled. So, one could vividly image that some of these bishops, when musing on the manner in which the logos (the Son) had been instrumental in the organization of the world, would have had recourse to insights taken from the Stoa or from Middle Platonism or Plotinus. In doing so, they would have associated the Son, "through whose mediation all things were made," with the Platonic world of Ideas or with the logos of Stoic cosmology whose activity coincides with that of the World Soul. They must, in a sense, have identified the Son with these world-immanent principles, also when they acknowledged that the Son was entirely on the side of the Father. They, certainly, did not regard the world as an emanation from the One. Yet, in order to explain the Son's instrumentality in structuring the world, they had no other tools at hand than those offered by the Stoa and Middle Platonism/Plotinus: the logos (the *nous*) as the world-immanent ordering principle.

New Testament Sources

Already some late layers of the New Testament initiated a reflection on the logos' emanation from the Father as well as on the role the logos (the Word) assumed in the creation of the world. This is especially true in the Gospel of St. John, which opens with the verses: "When all things began, the Word already was. The Word dwelt with God, and what God was, the Word was. The Word, then, was with God at the beginning, and through him all things came to be (*panta di'autou egeneto*); no single thing was created without him" (Jn. 1: 1–4). With explicit reference to Wisdom literature, the Son or the Word is personified as God's beloved partner, who assists Him in calling forth all that exists: "Wisdom exults her noble origin. She dwells with God and the Lord loves her. She is initiated in the knowledge of God and takes part in all His works" (Wisdom, 8: 3–4). Exegetes point out that the term logos refers not only to the figure of Wisdom, but also to the world-immanent ordering principle in the tradition of the Stoa.

The idea that all things came to be through the mediation of the beloved Son is also evidenced in four (most probably deutero-) Pauline letters: the First Letter to the Corinthians, the Letter to the Colossians, the Letter to the Hebrews, and the Letter to the Ephesians. In the First Letter to the Corinthians we read: "For us there is one God, the Father, from whom all being comes (*ex hou ta panta*) and for whom we exist, and there is one Lord Jesus Christ, through whom all things came to be (*di'hou ta panta*), and through whom we exist" (I Cor 8: 6).

In the letter to the Colossians one may perceive traces of a polemic with a Jewish community engaged in Pythagorean mysticism.[10] The text reads as follows:

---

10   Richard DeMaris, *The Colossian Controversy. Wisdom in Dispute at Colossae*, in *Journal of the Study of the NT, Supplement series 96* (Sheffield: Academic Press, 1994), 134–45.

He (Christ) is the image of the invisible God, the first born (*proototokos*) who has primacy over all created things. In him everything in heaven and on earth was created (*en autooi ektisthè ta panta*), not only the things visible but also the invisible orders of thrones, sovereignties, authorities and powers. The whole universe has been created through him (*di'autou*) and in view of him (*eis auton*). He exists before everything, and all things are held together in him (*ta panta en autooi synestèken*). (Col. 1: 15–17)

The "invisible orders" clearly refer to cosmic powers. The Pythagoreans held that planets and stars were sacred representatives of the Cosmic Deity. To these representatives one gains access through insight into the numerical ratios that govern the music of the spheres (see above Plato's *Timaeus*). The particular type of Pythagorism to which the Letter to the Colossians alludes is, in fact, a Jewish variant: under the influence of Jewish angelology, hosts of angels are allocated to the orbiting celestial spheres. The Letter to the Colossians affirms that all these celestial hierarchies (*thronoi, kuriotètes, archai, exousiai*), are subordinated to the Son, who exists before them, and 'in whom' (*en hooi*) and 'through whom' (*di'hou*), all things in the world, the celestial angels included, have been created and are held together.

The Letter to the Hebrews, too, presents the Cosmic Christ as the one who is heir to the universe and who sustains the reality of all things: "He [God] has spoken to us in the Son whom he has made heir to the whole universe, and through whom he created all orders of existence (*di'hou kai epoièsen tous aioonas*): the Son who is the effulgence of God's splendor and the stamp of God's very being, and sustains the universe (*pheroon ta panta*) by his word of power" (Hebr. 1: 2–3). In the Letter to the Ephesians, finally, it is stated that God has made known his hidden purpose, namely "that the universe, all in heaven and on earth, might be brought into a unity in Christ, its head (*anakephalaioosasthai ta panta en tooi Christooi*)" (Eph. 1:10). On the whole, these 'Pauline' texts seem to indicate that the cosmic Christ enables the subordinate celestial powers (*tous aioonas*) to graciously carry out their cosmic functions, provided they acknowledge him as the head and summation of all cosmic operations.

Most exegetes agree that the above texts which all stress Christ's role in the creation of the universe are rather late layers of the New Testament. The theology of the Cosmic Christ cannot properly be ascribed to Saint Paul, since the texts on which this theology is based are deutero-Pauline in character. On the other hand, these texts are part of the canon, which allowed them to have a bearing on the formation of Christian theology in general, and on the Nicene Creed, in particular. They became a favorite text collection, because of the inherent logic they displayed: if Christ is to bring about salvation from God, he ought to share with God in his cosmic creative power. The mediator in the order of salvation must possess that mediating power also in the order of creation.

CHAPTER 5

Medieval Times

The more the Triune God's distinctiveness from the world is affirmed, the less reticence Christians had in borrowing elements from ancient Greek cosmology. The cosmic spheres came to be seen as manifestations of God's glory descending from the vault of the heavens. The same deutero-Pauline text with its polemic against "the invisible orders of thrones, sovereignties, authorities and powers" (Hebr. 1: 3) gradually came to be read as an affirmation of the existence of hosts of angels, subjected to Christ, that were busy passing on motion to the rotating spheres to which they were allotted. In the medieval perception, the hosts of angels exert their influence on earthly events, symbolize the Christian virtues, and receive the souls of the deceased into their heavenly domain: the purer the souls, the higher they will ascend into the celestial spheres.

Saint Bonaventure

Bonaventure (1217–1274) was a contemporary of Thomas Aquinas (1224–1274), and also an Italian. He entered the order of the Franciscans during his studies in Paris. There he lectured from 1250 till 1256, the year he became minister general of the Franciscans. In his writings the Aristotelian influence begins to break through, though he clearly opts for the mystical orientation of Neo-Platonism. Bonaventure is heavily influenced by the Gospel of Saint John and the role the latter attributes to Christ, the expressive Word of the Father. For Bonaventure, the ultimate truth lies in the Christian message, into which cosmological insights must be integrated.

Bonaventure's Cosmology

Bonaventure gives evidence of being acquainted with Greek cosmology. In his *Breviloquium*, II, 3,1 he states that

> the entire material world machine comprises a heavenly and an elementary nature. The heavenly nature is mainly divided into three major heavens: the empyrean, the crystalline heaven, and the firmament. Within the firmament (the starry heaven) are the seven planetary spheres: the spheres of Saturn, Jupiter, Mars, Sun, Venus, Mercury, and the Moon. The elementary nature [under the Moon] is divided into four spheres: fire, air, water, and earth. From the highest point in heaven to the center of the Earth there are in all ten celestial and four elementary spheres. Thus, the whole material world machine is constructed in a distinct, perfect, and ordered manner.[1]

In this description one easily recognizes Aristotle's distinction between the celestial spheres and the sublunary spheres that contain the four elements. Yet, what is striking is that it omits an account of Aristotle's more complicated "world machine" in which the number of spheres around the planets was increased from 7 to 55. Bonaventure is apparently not interested in the intricate difficulties in which Aristotle was involved when trying—in the wake of Eudoxus—to reduce the apparent irregular motions of the planets to a combination of regularly rotating circles. Instead, he calls attention to two supplementary heavens, the flaming, empyrean heaven and the blue, crystalline heaven, which are located between the sanctuary of the Triune God and the firmament with the fixed stars. So, Bonaventure must also have been engaged in critical dialogue with another, unnamed source.

This source is, most probably, the Arab neoplatonic philosopher Avicenna with whom Bonaventure enters into polemics when expounding his view on the role of the angels.[2] Avicenna adopts Plotinus' system of emanations: The first origin of all things is the One. From the One emanates the first intelligence which is the seat of the realm of ideas. From this intelligence emanates the second intelligence, the World Soul, which moves the firmament with the fixed stars. From this intelligence emanates the third intelligence which moves the sphere of Saturn. From this intelligence emanates the fourth intelligence which moves the sphere of Jupiter, etc., till the ninth intelligence has been brought forth which moves the sphere of the Moon. Typical of Avicenna

---

1   Quoted in Max Wildiers, *The Theologian and his Universe: Theology and Cosmology from the Middle Ages to the Present* (New York: The Seabury Press, 1982), 43 (translation modified).

2   Ibid., 46.

is that he attributes to the ninth intelligence the role of *intellectus agens*, i.e., of the ever active intelligence that illuminates the human beings' receptive intellect, so that these, too, may acquire trustable knowledge in the midst of a world of change. In line with an old tradition in antiquity Avicenna identifies the cosmic intelligences with angels.

Compared with Avicenna, Bonaventure's cosmology is explicitly Christian. As we will see in more detail below, he developed an original view on the Trinity: the Father is the source-deity: from him emanates the Word that expresses the depth of the Father, and in doing so contributes to the emergence of the Holy Spirit. The Holy Spirit is the hypostasis that fences off the sanctuary of the deity. Even if one were to call the Son (or the Word) the "first intelligence," and the Spirit the "second intelligence," one would be mistaken in ranging them among the cascade of cosmic intelligences that move the planetary spheres. The cosmic intelligences (or angels) are created powers that are separated by a gulf from the uncreated Triune God.

This explains Bonaventure's eagerness to fill the "no man's land" between the Uncreated and the created with two heavens—the flaming, empyrean heaven, and the blue, crystalline heaven—which are not instrumental in setting stars and planets in motion. The empyrean heaven (*coelum empyraeum*) which is closest to God and which shares in His immobility is as the Burning Bush that shields the sanctuary of the Triune God from profanation. To preclude any direct contact between this heaven and the firmament, another heaven, the crystalline heaven (*coelum cristallinum*) is interposed. The crystalline heaven is, so to speak, cleansed with water, and reminiscent of the "waters above" mentioned in the first creation narrative of Genesis. Only then the firmament (*coelum firmamenti*) comes with the fixed stars. It is explicitly stated that God himself, the prime mover—although not without the assistance of some created power (a class of angels) —directly moves the firmament: "God moves the first mobile sphere by means of a created force with which He cooperates directly" (II *Sent.* d. 14, p. I, a. 3, q. 1).[3] The three highest heavens are exclusively God's domain.

If one calls the firmament the first heaven, then the second and third heaven above it constitute the domain in which the Triune God affirms his self-distinction from the created reality. Bonaventure jumps to this conclusion on biblical grounds. Had not Saint Paul declared (2 Cor 12: 2) that he had been caught up as far as the third heaven (the empyrean heaven which is closest to the Almighty God)? To this sacred place no ordinary mortal being usually gets access. Yet, when it fell to Saint Paul to be transported in

---

3   Ibid., 46.

rapture to the empyrean heaven, this heaven must already have been populated—with angels. The highest heaven is the place where choirs of angels
chant the glory of the triune God. It is at this juncture that Bonaventure
departs from Avicenna, who only admits of the existence of as many angels
as were needed to move the spheres of the seven planets. For Bonaventure,
this restriction is unfounded, for God created myriads of angels: "The angels
were created in the first place to behold God, and, in subordinate order, to set
the planets in motion. This task is for that matter only temporary, because at
the end of time all motion of the planets will finally cease."[4] Bonaventure is of
the opinion that the world will come to an end at the moment when enough
blessed souls will have gone to heaven so as to fill up the empty seats of the
perfidious angels that were thrown into hell.

The cosmic angels that move the planetary spheres are, to be sure,
important, because the kinetic energy they pass on to these spheres has a
bearing on the earthly processes of generation, growth, and perishing. Yet,
one ought not to unduly extol them since they are not superior to the angels
God has entrusted with a specific task: "Just as it is fitting that angels are sent
for the benefit of man, it is no less appropriate that they are instructed to set
the celestial spheres in motion and to direct them" (II *Sent.* d. 14, p. I, a. 3, q.
2).[5] Moreover, the cosmic angels have a limited effect; they only influence the
physical realm of the humans. In addition to these angels, there are incompa-
rably more angels (think, for example, of the angel guardian) sent to prepare
human minds for receiving the imprint of divine illumination. This statement
is directed against Avicenna, who held that the angel allotted to the sphere
of the moon (the ninth intelligence) illuminates the human intellect. For
Bonaventure, this task of illuminating the mind belongs only to Christ, "the
light of the world' " (Jn. 8:12).

The Triune God: Christ the Expressive Word

From the outset, Bonaventure presents the eternal Son as the expressive Word of the Father. The Father is the source-deity from whom the Son
emanates, not only as the one who is of the same substance as the Father (see
the Nicene Creed) but as the one in and through whom the Father utters and
expresses the fullness of his divine life. The Word is the total self-utterance
and self-expression of the Father (*Verbum est expressio Patris*). In this approach
the whole focus is on the speech act, on the Father's self-clarification and
self-presentation. Most probably Bonaventure came to this insight through a

---

4   Ibid.
5   Ibid.

meditation on the opening verse of the Johannine Gospel: "In the beginning was the word" (*in principio erat verbum*). The term "beginning" mentioned in this verse can—in a more philosophical context—be understood not as referring to a beginning in time but as referring to the Father, to the source-deity prior to which there is no beginning: thus to the "first principle" (*principium*). "In the beginning was the word" would then mean: from all eternity, the revelatory word was in the source-deity. From all eternity the Father stood, so to speak, in need of speaking to himself the word that allowed him to grasp the whole fecundity of his own being. "Through the Word that comes forth from him the Father utters himself and all things."[6]

In their reflection on the Trinity the Latin theologians focus on the distinct divine persons in the Trinity, whereas the Greek theologians emphasize the central position of the source-deity from which the Son and the Spirit emanate. Bonaventure remains faithful to the Greek-Byzantine tradition, while at the same linking the emanation process to the source-deity's speech-act through which the hidden ground of the source-deity expresses itself in the Word. This not only results in the active self-presentation of the source-deity, but also in the emergence of a second person in whom the whole force of this active self-presentation resonates. For Bonaventure, the Word expresses the depth of the source-deity not only to the source itself, but also to others: in this case, to and in the world "outside" of the deity.[7]

In this function, the Word communicates its expressive force to the things created. These begin to imitate the theo-expressivity of the Word in their own domain, since once called into existence they are empowered to share in the expressive force of the Word. In terms of causality this means that the Word (the Son) is the model, the exemplar (*causa exemplaris*) to be imitated by the whole of creation. The created entities, too, are endowed with the capacity to give expression—in an analogous way—to the source-deity who through them seeks to come to visibility. In activating this capacity the created entities "imitate" the Word's theo-expressive force in the created realm. Let us note that on the basis of this approach Bonaventure will be able to show that the "expressive Word" continues to give expression to the source-deity in a human form, as the "Word incarnate."

---

6   Bonaventure, I *Sent.*, d. 32, a. 1, q. 1, fund 5: Pater verbo suo quod ab ipso procedit dicit se et omnia. See Jacques-Guy Bougerol (ed.), *Lexique Saint Bonaventure* (Paris: Editions Franciscaines, 1969), 83.

7   See Bonaventure, I *Sent.* d. 18 a.un. q. 5 ad 4 (Quaracchi I 331b). Quoted in Wilfried Schachten, *Intellectus Verbi, Die Erkenntnis im Mitvollzug des Wortes nach Bonaventura* (Freiburg/München: Karl Alber, 1973), 119.

The Word or the Son is the canon of all beauty because his expressive life is fully transparent towards the source-deity who comes to light in him and "from whom he never departs."[8] The way he relates to the source-deity will become normative for the behavior of all things created. The exemplarity of the Son entails a deontology for created existence. So, it is important to analyze why the Son deems it necessary to "give back" his beauty to the original giver. This analysis reads as follows: (i) the Father is the inexhaustible source-deity (*fontalitas*). (ii) In this quality, He cannot possibly abdicate his position of being the source, but (iii) He can empower another divine person whom He begets from all eternity to fully share in his life-force and to give expression to it. For Bonaventure, the Father possesses the fullness of life. It is this fullness of life that is passed on to the Son, so that it may become entirely visible in him. Bonaventure refers to the Johannine dictum: "As the Father has the plenitude of life in himself, so has the Son, by the Father's gift" (Jn 5: 26).[9] The Son thus expresses in his own being the source-deity's plenitude of life, although he is not the origin of it. (iv) The Son continually lives in a state of amazement and praise, realizing that the life-force He activates is simply the source-deity's plenitude of life. This amazement leads him to a life of thanksgiving and glorification, returning his own glory to the Father. This rendering back of the glory can rightly be termed: the bringing forth of the Holy Spirit, who proceeds from and vivifies the Son's act of thanksgiving— the Holy Spirit, who in the Latin version proceeds from the Father and the Son (*filioque*).

The beauty of the Son consists in his being permanently "turned back to the Father, especially at the moment when he feels he activates within himself the life impulse of the source-deity. Again Bonaventure takes his inspiration from the Gospel of Saint John: "God's only begotten one, who is nearest to the Father's heart—*ho oon eis ton kolpon tou patros*—he has made him known" (Jn 1: 18) . This Johannine influence basically entails a reversal of the scheme of the *Timaeus*, where the Demiurge fixed his gaze on the eternal ideas (or in Greek patristics: on the Son as the dwelling place of the ideas). In Bonaventure's vision, the Son or the expressive Word fixes his gaze on the Father, as on the unfathomable source-Deity, whose whole life-potential he brings to visibility. In doing so, he is able to give expression to the fullness

---

8   Peter Lombardus, 1 *Sent.* d. 3, pars 1, cap 1. Quoted in Hans Urs von Balthasar, *The Glory of the Lord*, Vol. II *Studies in Theological Style. Clerical Styles*, trans. Andrew Louth e.a. (Edinburgh: T & T Clark, 1984), 300.

9   See Bonaventure, *Quest. Disp. De Myst. Trin.*, q.8 ad 7. Quoted in Gilles Emery, *La trinité créatrice* (Paris: Vrin, 1995), 183.

of divine life, as well as—derivatively—to the ideas that will be formative of the things created.

Theo-expressivity of the Created Realm

Bonaventure's religious vision of the world has been an inspiration for many generations. For him, the cosmos as a whole has been created in the image of the Son. It, therefore, shares in the Son's expressive force. All intellectual creatures (including the angels) seem to be endowed with an original expressive quality. They express themselves and the hidden ground that sustains them. At their own level of participation, they display a splendor that is, at once, both totally theirs and pure gift from the hidden deity. So, they feel the need to reflect their glory back into its generous origin. Yet, it is precisely at this juncture that they are exposed to an inner tension. They may yield to the temptation of clinging to their own beauty, thus ignoring that it is God-given. This is the temptation to which Lucifer succumbed. The light-bringer became a fallen angel by adoring his own brilliance without recognizing the source-deity as his animating origin. Bonaventure dwells at length on the fallen angels' expulsion from heaven.[10] On the other hand, he regards this expulsion as a fact of the past. The choirs of angels that chant the glory of the triune God as well as those who move the celestial spheres have resisted the lure of self-adoration: they all give their glory back to the Creator, after the model of the eternal Son.

The contemplation of stars and planets provides us with a potent incentive in our ascent to God (*Itinerarium Mentis in Deum*). The angels who move the planetary spheres are objects of admiration because they visibly act as second causes giving back their glory to the heavenly originator. What they are doing serves as a model for the humans in the sublunary sphere. On their lower rank in the hierarchy of being, the humans, too, are called to uniquely express their own beauty without falling prey to self-adoration. The kingdom of God increasingly attains its fullness as humans increasingly succeed in expressing their intrinsic worth, and referring it back to the glorious Giver.

Bonaventure extends this religious vision of the world to the subhuman domain: to animals, plants and inorganic entities. For him, the woods and clouds and animal species on earth are worthy of praise because they are echoing the expressive process in which the Word brings the source-deity to light. In their particular mode of existence, rivers, roses, and birds display

---

10   See e.g., Bonaventure, *Hexaemeron*, I, 17 (Quaracchi V, 32). Quoted in Wilfried Schachten, *Intellectus Verbi*, 148.

their own expressive force while, at the same time, bearing witness to the hidden ground that sustains them.

Bonaventure borrows from the Neoplatonic theory of participation (Plotinus, Proclus, Dionysius) as many elements as possible to account for the God-given splendor of the particular entities. He makes it clear that the real meaning of the "One" is: the "Unique one," and that all the particular entities the One calls forth must, according to various degrees, partake in this uniqueness. Each and all of them enjoy an unrepeatable, irreplaceable splendor that is linked to their particular self-expression. The Jesuit Gerald Manley Hopkins (1844–1889), who became acquainted with Bonaventure's religious vision of the world through his study of the Scottish Franciscan Duns Scotus (c. 1265–1308), wrote several poems to celebrate this remarkable phenomenon. In these poems, he honors the baffling uniqueness of each entity's expressive force —a uniqueness that evokes the "wild" song of the eruptive ground of the Godhead.

> All things counter, original, spare, strange.
> We see the glories of the earth,
> But not the hand that wrought them all
> .....
> Give beauty back, beauty, beauty, beauty,
> back to God,
> beauty's self and beauty's giver.[11]

In one of his *"Quaestiones"* Bonaventure asks the question as to whether it would not have been better for humans to enjoy incorruptible existence in the sublunary sphere. The answer is negative, for in that case humans would lack the capacity to express divine splendor through the humility of human flesh. This view is inspired by the incarnation, in which God's Word took on flesh as a privileged medium of expression. It also indicates Bonaventure's appraisal of the earth. For him, humility (*humilitas*) is associated with soil/earth (*humus*). This earthly makeup makes it easier to glorify the Creator, especially when observing how even the most humble earthly entities are endowed with an expressive force through which they display their own numinous core and that of the Creator. In spite of his Neoplatonic inspiration, Bonaventure developed a positive appreciation of the senses. No genuine esthetics can exist in the sublunary sphere without

---

11   *The Poems of Gerard Manley Hopkins*, 4th edition, ed. W.H. Gardner (Oxford, 1956), 22, 37, 59.

the use of the senses. Sensory feelings are essential to give expression to one's inner expressive core and to glorify the Creator.

Thomas Aquinas

In his mature period Thomas Aquinas (1225–1274) not only wrote his *Summa Theologiae*, but also commentaries on Aristotle's *Physics* and *Metaphysics*. He found this intense study of Aristotle necessary in order to refute some of the interpretations that the Arab philosophers Avicenna and Averroes had given of his works. Aquinas sought to understand Aristotle better than the Averroistic school in Paris, with a view of transposing the current Augustinian theology into the key of Aristotelism.

The Question of the Eternity of the "World"

Aquinas' commentaries on Aristotle give a good picture of the philosopher's basic categories: the transition from potentiality (*potentia*) to actuality (*actus*), the theory of motion, the difference between the sublunary realm and the celestial spheres, and the impact of the celestial motions on the continuity of life on earth. There is, however, a specific point he endeavored to clarify. In their reading of Aristotle, the Parisian Averroists maintained that the 55 cosmic intelligences were as eternal as the Prime Mover. Bonaventure had already tackled this problem, and pointed out that revelation teaches us that all of the heavenly powers, other than God himself, have a beginning in time. For Aquinas, however, the dispute ought to be settled first on philosophical grounds.

The dispute with the Averroists must be placed in the broader context of their theory of the "double truth." According to this theory, there is an irreconcilable clash between truths of faith and truths arrived at by means of reason, so that any attempt at harmonizing both is useless. The theory of the double truth was condemned by the Archbishop of Paris in 1270 and again in 1277. Thomas Aquinas, who studied in Paris from 1245 till 1248 and from 1252 till 1256, knew about these polemics. In 1269, one year before the condemnation, he was even called from Italy to Paris to settle the conflict. He realized what was at stake: the beginning of a secularized culture, emancipated from its Christian heritage. Aquinas' synthesis sought to avoid the cleavage between philosophy and faith. He was convinced that, since there is only one God, the origin of all things, *both* the philosophical approach *and* the theological approach to truth, if assiduously pursued according to their own methods, cannot basically contradict each other.

Aquinas wrote a special essay *On the Eternity of the World,* that is: on the eternity of the cosmic celestial machinery, but his clearest elucidation of this topic is to be found in the *Third Way* to prove the existence of God in the *Summa Theologiae*. The proof of God's existence in this *Third Way*, divided into two parts reads as follows:

> The third way is based on what need not be and what must be, and runs as follows. (i) Some of the things we come across can be but need not be, for we find them springing up and dying away, thus sometimes in being and sometimes not. Now everything cannot be like this, for a thing that need not be, once was not; and if everything need not be, once upon a time there was nothing. [...] Not everything therefore is the sort of thing that need not be; there has got to be something that must be. (ii) Now a thing that must be, may or may not owe this necessity to something else. But just as we must stop somewhere in a series of causes, so also in the series of things which must be and owe this to other things. One is forced therefore to suppose something which must be, and owes this to no other thing than itself; indeed it itself is the cause that other things must be [...] This all men speak of as God.[12]

In the first part of the argument Aquinas starts from the sublunary entities; they are to be qualified as perishable, contingent entities: it is possible for them to be or not to be. What Aquinas has in mind are the entities that began to exist through natural generation and will cease to exist through natural perishing. The fact that they exist in this contingent way must be attributed to the steering effect of cosmic entities whose existence is necessary, i.e., by entities for which it is impossible not to exist, and which are thus eternal. Aquinas follows in big strokes Aristotle's proof of God's existence in his work *Physics*. This proof, too, started from considerations in the sublunary domain, such as the transition from potentiality to actuality and the transmission of the genetic code from generation to generation, phenomena that are steered by cosmic motions. Such a view rules out that the contingent entities would have begun by pure chance, or what boils down to the same: would have originated from nothingness, which in Aquinas' eyes is absurd. The only remaining alternative is that the contingent entities owe their existence to a necessary substance or to a series of necessary substances that have caused them to exist.

Only in the second part of the argument does Aquinas claim to strictly prove the existence of God, the first Origin, by pointing out that the necessary substances to which the contingent things owe their existence must,

---

12   Thomas Aquinas, *Summa Theologiae*, Volume 2: *Existence and Nature of God*, trans. Timothy McDermotte (Cambridge: Cambridge University Press, 2006), 15 (l. I, q. 2, art 3).

in turn, owe their necessary existence to a First Necessary Being that has of itself the necessity to exist. To grasp the point of this argument, one has to bring to mind Aristotle's cosmology in which the planetary spheres are moved by cosmic intelligences. These cosmic intelligences—as well as their respective spheres and planets—are indestructible and not subject to generation and decay, as is the case in the sublunary domain. What is more, these cosmic intelligences are, just as the Prime Mover who sets the firmament in motion, without any spatial extension. This implies that they specifically share in God's everlasting eternity forward and backward: they are as imperishable and eternal as the Prime Mover.

Confronting this state of affairs, Aquinas draws a decisive distinction between a necessary (eternal) substance having *of itself* the necessity to exist and necessary (eternal) substances having their necessity to exist caused by another, i.e., by a First Cause that places them in their position of eternal immaterial substances. This distinction highlights the sovereignty of the First Cause. Only the First Cause possesses *of itself* the necessity to exist—and is therefore called God—whereas all the other imperishable substances receive their necessity to exist from the First Cause. To corroborate this insight Aquinas, just like Aristotle did in his proof of God's existence, demonstrates the impossibility of an infinite regress. Even if one admits of a series of necessary substances that receive their necessity to exist from another substance, such a series cannot go on to infinity; it must terminate in an independent initiator who has the necessity to exist *of itself.* "But just as we must stop somewhere in a series of causes, so also in the series of things which must be and owe this to other things. One is forced therefore to suppose something which must be, and owes this to no other thing than itself."

For modern readers, it is difficult to visualize the causal dependence of the eternal cosmic intelligences on the equally eternal First Cause. Not so, for Aquinas, who made ample use of the *Liber de Causis* (Book on the Causes), a compendium of Neoplatonic tenets in which the Plotinian notion of participation was explained. In our analysis of Plotinus it became evident that the One produces everlastingly, and that its products are less than Itself. Nothing can come from the One which is greater than the One. The One remains elevated above the entities that emanate from It in an a-temporal manner. The Plotinian emanations do not have a beginning in time; they take place from eternity to eternity. But even then—without temporal beginning—they are and remain causally dependent on the One, as the simplest origin. This is the message Aquinas conveys with his distinction between a necessary (eternal) being having *of itself* the necessity to exist and necessary (eternal) beings receiving their necessity to exist from another. His reply to the Averroists is clear: also when the cosmic intelligences are as eternal as God, they remain dependent on the First Cause from which they receive their eternal existence. Moreover, this dependence forms, so to speak, the transition to a dependence that includes a temporal beginning.

Indeed, for Aquinas, it is only on biblical grounds that we know that the whole celestial realm, the cosmic intelligences included, had a beginning in time. In his *Commentary on Aristotle's On the Heavens* he writes:

> We do not say according to the Catholic faith that the heavens always existed, although we say that they will endure forever. Nor is this against Aristotle's demonstration here, for we do not say that they began to be through generation, but through an efflux from the first principle, by whom is perfected the entire existence of all things [...] We posit that the heavens were produced by God according to their whole substance at some definite beginning of time.[13]

For Aquinas, revelation teaches us that God calls forth all things out of nothing. For the cosmic intelligences and the whole celestial realm this means that, indestructible and everlasting as they are, they had a real beginning in time. But he is quick to add that this beginning differs from the generation of perishable entities in the sublunary domain. Their beginning in time bestows on them a necessary existence that will, from that moment on, last forever.

In his *Summa Theologiae*, Aquinas dwells at length on the properties of the angels, i.e., of the Christianized cosmic intelligences. He makes it clear that once they were created, the angels enjoy an everlasting, indestructible existence. But since we have to do with an everlastingness that had a beginning in time, the angels totally differ from the only begotten Son of God, who was begotten before the beginning of time. That is the reason why the eternal time span of the angels is called *aevum*, an everlasting time span that had a beginning. Only the Triune God is the everlasting Eternal One (*sempiternus*). Let us not forget that for Aquinas also the souls of the deceased will last forever, although they were created in time. Finally, in *Summa Theologiae* I, q. 46, art. 2, Aquinas asks himself the question as to whether natural reason can conclusively prove that the cosmos had a beginning in time. His answer is negative: "That the world (*mundus*) has not always existed," he writes, "cannot be demonstratively proved, but is held by faith alone. [...] The origination of the world cannot receive any consistent explication from within

---

13   Thomas Aquinas, *On the Heavens*, trans. Fabian R. Larcher and Pierre H. Conway, I, Lect. 6, n. 64 http://dhspriory.org/thomas/DeCoelo.htm#1-1 (accessed March 30, 2011).

the world itself"[14] Only God can reveal to us the mighty deeds he performed in creating the world; it is from this revealed truth that theological reasoning deduces that the "world" had a beginning in time.

Aristotle or Ptolemy?

Aquinas' considerations about the dependence of necessary (eternal) cosmic entities on the eternal Creator are so balanced and classic that they transcend any particular type of cosmology. On the other hand, it is evident that with regard to his doctrine of God, Aquinas heavily relied on Aristotle's theory of motion, which, compared to what we now know, is totally outdated. In this light it is worth examining his assessment of Ptolemy's achievement.

Ptolemy's System of the World

Ptolemy (ca 85–165) was an astronomer and geographer who lived in Alexandria and wrote in Greek. He is the author of the famous *Geography*, which remained the standard geographical work throughout the Middle Ages, and of the *Great Collection* or *Almagest,* in which he gave a full account of Greek astronomy including his own system. In order to explain the particular motions of the planets on their path along the ecliptic, Ptolemy deemed it necessary to abandon the clusters of concentric spheres that had been used by Eudoxus, Callippus, and Aristotle. According to Ptolemy, these clusters could not shed any precise light on the chief peculiarity of the wandering stars (planets), i.e., on their looping. When observed against the background of the fixed stars, the planets, except the Sun and the Moon, seem, at regular intervals, to halt, adopt a retrograde motion, and the resume their regular direct motion.

Compared to his predecessors, Ptolemy took a significant step forward. For him, the assumption of two *concentric* interacting circles, one inside the other, was totally inadequate to explain the looping of the planets. What is needed is a non-concentric construction in which a large rotating circle, termed the deferent, interacts with a smaller rotating circle, termed the epicycle, mounted upon the former. If you move the deferent anticlockwise, the epicycle will also move anticlockwise but at a higher speed because of its shorter radius. And further: if you put a planet on the circumference of the epicycle, and you look at the rotating epicycle from the Earth in the middle, you will see the planet move forward as long as it is on the section of the

---

14   Thomas Aquinas, *Summa theologiae*, Volume 8: *Creation, Variety and Evil,* trans. Thomas Gilby (Cambridge: Cambridge University Press, 2006), 69 (I, q. 46, art. 2).

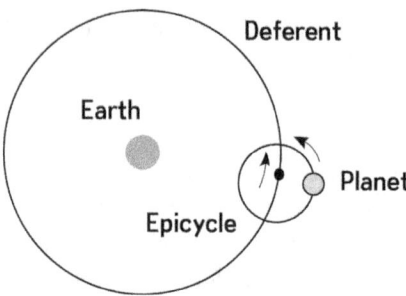

Figure 5. Epicycle

epicycle's circumference that is outside the rotating deferent. But as soon as the segment of the epicycle's circumference on which the planet is attached enters the inner side of the rotating deferent, it is seen as halting and taking on a retrograde motion. Once the planet on the epicycle's circumference turns again outside the rotating deferent, it resumes its direct regular motion.

To refine this system, later astronomers mounted even smaller epicycles upon the original ones, and still other smaller ones upon those newly added epicycles, etc., in order to obtain additional rolling effects. This accretion of epicycles was in principle open-ended, and made some lament as to why the Creator had devised such an intricate plan for the world system.

Ptolemy's solution was, without doubt, superior to that of Eudoxus and Aristotle. Still, his basic concern was pretty much the same, viz., to explain the apparently irregular motions of the planets through a combination of regularly rotating circles, the circular rotation being regarded as the most perfect motion in the universe. Ptolemy, however, opted for a totally different arrangement of the rotating circles. He gave up the principle that all these circles must have their center in the center of the universe. That is why his system is called the system of the *eccentric* spheres. It is eccentric, i.e. non-concentric, for two reasons. First, the epicycle has its center in a point on the circumference of the deferent. Second, the deferent need not have its center in the center of the universe, the only requirement being that the center of the universe must be somewhere inside the deferent. Moreover, for the different planets, the size of their respective deferent circles as well as of their epicycle(s) could vary considerably.

The accretion of epicycles eventually grew so complicated that it aroused the suspicion of many scientists. At various occasions astronomers and geometers would venture into simplifying the system. The decisive attempt came from Copernicus, who showed that it was much more consistent to have the planets' deferents and epicycles orbit the Sun, rather than the Earth. As we will see below, Copernicus came to this insight because Ptolemy had already changed the order of the planets so as to assign a more central place to the Sun. Counting from Saturn on, the Sun now occupies the fourth place, instead of the sixth in Aristotle's classification, thus coming practically in the middle.

Aquinas' Assessment of Ptolemy

Aquinas acknowledged Ptolemy's merit: his system of non-concentric deferents and epicycles makes it possible to understand the mechanism of the looping of planets. But he wondered why the reconstruction of the peculiar motions of Mercury and the Moon was so complicated. For Mercury, Ptolemy had to postulate that "the center of its deferent is moved in a small circle about the center of the world."[15] And to explain the Moon's receding nodal points, a whole array of epicycles was needed. For the Sun, however, things suddenly became much simpler: this planet did not even need any epicycle.

When pondering these results, Aquinas opined that Ptolemy had not yet resolved the intricate problem of the heavenly ballet. Why must Mercury, and the Moon, the planets closest to us, have so many movements, whereas the Sun, which is in the middle of the series, has so few? "Mercury and the Moon, the lowest of the planets, have the most motions, whereas the Sun, which they place as intermediate, has the fewest, with the remaining planets being in between."[16] From these observations Copernicus will later draw the conclusion that the Sun must be placed in the center of the universe, and that the Earth and the planets orbit this center. Aquinas reached a totally different conclusion. For him, Ptolemy's explanation of the planetary motions was not any better than that of Aristotle: they were both conjectural hypotheses, having their pros and cons.

Aquinas gives evidence of having a fairly good knowledge of the systems of Eudoxus, Callippus, and Aristotle; he even mentions the reasons why Callippus and Aristotle augmented the number of the concentric spheres from 26 to 33 and 55. In his commentaries on Aristotle's *Metaphysics* and *On the Heavens* he gives a detailed review of this "ancient school," whose results he compares with Ptolemy's system. Yet, towards the end of his *Commentary on the Metaphysics,* Aquinas is so much under the spell of Aristotle's picture of the world that he uses it to expose the alleged weaknesses in Ptolemy's system. He points out that the basic presuppositions of

---

15   Thomas Aquinas, *On the Heavens*, II lect. 17, n. 454.

16   Ibid.

this system "are contrary to the truths established in [Aristotle's] philosophy of nature."[17]

A first objection relates to the eccentric circles. If this assumption is accepted, then "not every motion will be either towards or away from or around the center of the world."[18] In other words, Ptolemy's system militates against the basic principles of the Aristotelian theory of motion. Similar objections are made against the use of epicycles: "From the hypothesis of epicycles it follows either that the sphere by which the epicycle is moved is not whole and continuous, or that it is divisible, expansible, and compressible in the way in which air is divided, expanded, and compressed when a body is moved."[19] In other words, the continuous circularity of the deferent is endangered when an epicycle is mounted upon it, which is, again, at odds with Aristotle's high esteem of the perfect circular motions in the quintessential realm.

In this assessment Aquinas puts his cards on the table. He no longer suggests, as he had done earlier, that the astronomic systems of Aristotle and Ptolemy are on the same footing because of their hypothetical character. The accusation is straightforward: Ptolemy offends against the basic tenets of Aristotelian physics. In voicing this accusation Aquinas reveals the extent to which he espoused the basic principles of the "Philosopher." Had he dared to call them into question, this would have shaken the foundations upon which rested his theology of the First Unmoved Mover. Let us note in passing that for Aquinas only theology and philosophy provide necessary truths, whereas other sciences, like astronomy, basically work with hypothetical assumptions.

---

17   Thomas Aquinas, *Commentary on the Metaphysics*, trans. John P. Rowan (Chicago: Henry Regnery Company, 1961), 904, n. 2570.

18   Ibid. n. 2568.

19   Ibid, n. 2569.

CHAPTER 6

The Copernican Turn and its Confirmation by Kepler and Galileo

End of the Religiosity of the Cosmic Spheres?

The religiosity of the cosmic spheres, with its sharp demarcation between the eternal heavenly motions and short-lived existence in the sublunary realm, continued to exert an enormous fascination on the Christian milieu. Although the Aristotelian understanding of God has little in common with the Christian ideal, the heavenly ballet he seemed to initiate had enormous symbolic significance for the people's aspiration for personal immortality. In their study of ancient cosmology Merleau-Ponty and Morando capture this cosmic vision:

> Are we to call this Prime Mover God? Yes, but such a God is quite different from the Christian God. First, he is not the only God, and second, he presides over the order of the world without being concerned about its souls. If he seems to act upon them, he does so as the model of a perennial harmonious existence that offers itself for their admiration. Yet, however suspect this doctrine may appear, the cosmology of the spheres had the advantage of responding to a certain expectation of common sense. Since humans are basically exposed, albeit in various degrees, to the bad fortune of sickness, famine, war and death, the spectacle of the celestial motions, indifferent to the "tumult and furor" of the Earth, and infinitely regular, spontaneously evoked the idea of an existence freed of the fetters of decline: "up there" was to be found only order and beauty.[1]

---

1   Jacques Merleau-Ponty and Bruno Morando, *Les trois étappes de la cosmologie* (Paris: Laffont, 1971), 49 (translation mine).

Besides Plato's musical presentation of the World Soul whose mathematical proportions were visible in the rotations of stars and planets,
Aristotle's doctrine of the first mover and the cosmic intelligences that
moved the firmament and the planetary spheres was another classic
expression of Greek *theo-cosmology*. Its influence on ancient civilization
and on Greek and Latin Christian culture cannot be underestimated. True,
the Christian milieu placed a heavier emphasis on God's sovereign act of
creation, but on the whole, a symbiosis had taken place between the cosmic
God and the God of Jesus Christ, as we have seen in our presentation of
Bonaventure and Aquinas. Not only did the Christian milieu pay respect to
the sacredness of the cosmos, it also assimilated this heavenly orientation
into its liturgical texts. In both the Byzantine and Latin liturgy the bestowal
of God's grace is celebrated as a cosmic event in which the *triune* God's
immortal beauty descends from heavens unto earth. This helps us understand the extent to which Christian believers found a deep inspiration in
Ptolemy's epigram:

> When I—though mortal and impermanent—am allowed to contemplate
> the starry sky at night, I feel myself no longer dwelling on earth,
> but close to the Creator. Then my spirit drinks from the source of Eternity.[2]

For centuries, the religiosity of the cosmic God, visibly perceptible in
the celestial rotations, had such a potent cultural impact that great innova-
tors like Copernicus (1473–1543) and Kepler (1571–1630) were at pains to
abandon it. Their worry did not come so much from the fact that it was
proven now that the Earth orbited the Sun and not the other way around.
They were rather alarmed by the realization that the neat separation between
the celestial spheres and the sublunary domain could no longer be upheld. It
began to dawn upon them that there were no special criteria of perfection
reserved to the celestial spheres and their cosmic intelligences. It became
obvious that stars, spheres, and planets are not made of a specific quintessence. Their building blocks, like those of everything else, turned out to be
mere matter, matter that obeys the universal laws of physics. The beautiful
construct of the perennial heavenly rotations lacked any solid foundation.

Gradually the splendor of cosmic liturgy was going to be consigned to the
past. To the extent that the impressive liturgical ceremonies of the Orthodox

---

2   Franz Boll, *Das Epigramm des Claudius Ptolemaeus* in Id, *Kleine Schriften zur Sternkunde
     des Altertums*, ed. V. Stegeman and E. Boer (Leipzig, 1950), 143 (translation mine).

and the Latin Churches feed on that cosmic awareness, the weakening of their charisma can already be predicted. Cosmology will be practiced now by a generation of scientists whose achievements are to lead to the birth of modern mechanics, and this mechanics hastens the downfall of the old religiosity. The law stating that a body endeavors to preserve its present state, whether it is a state of rest or of moving uniformly forward in a straight line, applies to both heaven and earth. One no longer stands in need of the existence of a host of cosmic intelligences whose seduction keeps the celestial spheres going. Farewell, cosmic movers and godly spheres in heavens.

Copernicus

The Copernican revolution took place in a modest way, and was carried out mainly in the domain of geometrical calculations. Nicolaus Copernicus, a Polish cleric and canon (1473–1543), arrived at his conclusions not because he was challenged by new astronomical data made available by telescopes. He still had to work with the astronomical tables that enabled calculation of the positions of the planets for any given time based on the Ptolemaic theory that assumed Earth to be at the centre of the universe. Rather, his restless mind sought to simplify the Ptolemaic theory by changing the centre around which the system of deferents and epicycles rotated. Copernicus' new centre would be the Sun.

In order to substantiate this new view Copernicus had recourse to deviant cosmological opinions of the past. In his *De revolutionibus orbium caelestium* (1531), he refers back to Plato, who in spite of his preference for uniform circular rotations had attributed a light rocking to the Earth in the middle of the universe. He also highlighted that, centuries before Ptolemy, Aristarchos of Samos (310–205) had already suggested how much simpler it would be to place the Sun in the middle of the universe with the Earth and the other planets revolving about it. In those days, however, this view met with serious opposition, the main objections being that a revolving Earth would unleash severe storms on oceans and seas and make ships fly out of the waters.

Taking Aristarchos' intuition as a lead, Copernicus began to elaborate his various axioms which can be summarized as follows.[3] He made it clear (first axiom) that if the Earth was spherical as was commonly believed even in earlier days, it ought to have a rotation about its own axis, just like the other planets, for it is evident that *the axial rotation naturally flows from the*

---

3   In his *Commentariolus*, written about 1514, Copernicus enumerates seven axioms. See Alexandre Koyré, *The Astronomical Revolution: Copernicus–Kepler–Borelli*, trans. R. Maddison (London: Methuen, 1980), 26–27.

spherical shape. So, whenever there is a spherical figure, it rotates around its axis: "each and every spherical body must of necessity carry out a *uniform* rotation about one of its diameters."[4] This can further be specified: the Earth rotates about its axis from the West to the East in 24 hours. This statement was the result of a daring thought-experiment that departed from the ancient presupposition of an Earth resting on a fixed location.

No wonder that this position met with the disapproval of the Aristotelians. They objected that it was impossible for the Earth to spin about its axis since such would give it a uniform circular movement, and thus a perfection that according to Aristotelian principles was not to be found in the sublunary domain. Moreover, it was argued, if the Earth were to spin that forcibly, then the clouds and the birds in the sky would have problems in catching up with its speed; if the birds would fly in the same direction in which the Earth were moving, they would eventually be left behind and even appear to fly backward, yet such a curious phenomenon was never observed. Furthermore, what might happen to the Earth itself? Was it not to be feared that, spinning at such speed, our planet would get out of joint, burst, and finally fall apart? Not bewildered at all, Copernicus retorted by asking whether they themselves did not fear that the highest firmament, which was believed to carry out a full rotation in 24 hours, might collapse under the tremors of such breathtaking speed. The Aristotelians were perplexed at this reply. For them, the heavens were made of eternal quintessence; so, in their mind, kinetic disasters simply could not happen "up there." Copernicus' opponents were apparently not able to grasp the very reason why the spinning Earth is not falling apart—which is also the reason why the Earth-dwellers do not feel that spinning—namely the selfsame uniformity with which the axial rotation is carried out: "a spherical body must by nature perform a *uniform* rotation on one of its diameters."

But Copernicus still made a further point. If one takes it for granted that all the planets, in virtue of their spherical shape, are endowed with an axial rotation, and if one further admits, in line with ancient cosmology, that they all have circular trajectories in space, why should one not conclude then that the Earth, with its axial rotation, must also of necessity follow a circular trajectory in space? *Axial rotation and locomotion in space go hand in hand* (second axiom). In other words, the fact that the Earth spins about its axis from the West to the East makes it embark on a circular trajectory—from the East to the West—around a fixed center. And what is true for the Earth is also

---

4   E.J. Dijksterhuis, *The Mechanization of the World Picture,* trans. C. Dikshoorn (New York: Oxford University Press, 1969), 289.

true for the planets: they, too, must adopt a circular trajectory about a middle point that carries them from the East to the West.

When, finally, asking himself which is the most plausible center about which the Earth and the planets rotate, the answer became obvious: it must be the Sun (third axiom). Why else did Ptolemy in his days find it necessary to change the order of the planets and to place the Sun exactly in the middle, i.e., between Mars and Venus and no longer, as in previous classifications, between Mercury and the moon? This centrality in terms of order made more sense, and yielded better results in accounting for the planets' composite motions within a system of rotating circles and epicycles. This recourse to Ptolemy, however, was only meant to overthrow his system. One must go further than Ptolemy and posit that the Sun is really the—searched-for—center about which all the rest revolves. "The hypothesis that the Earth orbits the Sun once a year, and that the planets Mercury, Venus, Mars, Jupiter and Saturn, each in their own periodic time of revolution, are doing the same, gives a picture of the cosmos that is both simpler and more harmonious than Ptolemy's."[5] Such an increase in simplicity and harmony brings us closer to the truth.

Copernicus' hypothesis is, indeed, much simpler. It has two major advantages. First, it frees the firmament and the fixed stars of their breathtaking

Aristotle	Ptolemy	Copernicus
Saturn	Saturn	Saturn
Jupiter	Jupiter	Jupiter
Mars	Mars	Mars
Venus	Sun (middle position)	Earth (plus Moon)
Mercury	Venus	Venus
Sun	Mercury	Mercury
Moon	Moon	Sun (heliocentric)
Earth (geocentric)	Earth (geocentric)	

Figure 6. Order of the planets according to Aristotle, Ptolemy, and Copernicus

daily rotation, for in the heliocentric approach, the distant firmament is in a relative state of rest. The firmament's supposed daily rotation is in fact nothing else but the *optical projection* on to the heavens of the only decisive rotation that in the meantime takes place: the daily axial rotation of the Earth. Because once in 24 hours the Earth rotates eastward about its axis, the observers on Earth get the impression that in that period of time the whole firmament revolves westward about the Earth. One ought also not to forget

---

5   Ibid., 322.

that the firmament is much farther away from the planets than previously assumed: "the distance separating the Earth from the Sun is insignificant compared with that separating the Sun from the sphere of the fixed stars."[6] (fourth axiom)

A second advantage of the Copernican revolution is that it makes all the planets except the Moon (which is supposed now to orbit the Earth) revolve about the Sun as their common center. All the planets take on a circular trajectory from East to West (counter-clockwise), each at a speed that decreases with their distances from the Sun. Each planet has its proper rhythm, its own year, to carry out its circular journey around the Sun. This, too, entails an optical debunking, this time of the apparent eastward journey of the planets along the ecliptic, which make them—in the old view—lag behind the westward journey of the firmament. Indeed, it takes Saturn 30 years to orbit the Sun, and Jupiter 12 years and Mars almost 2 years to complete their annual revolution. So, it is clear that viewed from the Earth which orbits the Sun much faster—in 365 days—these planets appear to lag behind on their westward journey around the Sun. It is because the observers on Earth view the other planets' journeys around the Sun *from the backdrop of their own annual revolution (of which they are ignorant)* that they "perceive" a lagging behind of the superior planets, those further away from the Sun than Earth. For the inferior planets, those closer to the Sun than Earth, i.e., Venus and Mercury, a similar discrepancy is perceived, since they, in turn, orbit faster around the Sun than the Earth.

Yet, a full account of the apparent path along the ecliptic is not yet given, though. What is still to be explained is the fact that the path along the ecliptic was perceived as forming an angle of 23.5° with the equator of the universe. From Copernicus we know that every 24 hours the Earth spins around once on its axis. But later scientific research showed that the Earth's axis is tilted 23.5°degrees with respect to the earth's orbit around the Sun. This tilted axis explains why the observers on earth get the impression that the apparent eastward journey of the planets follows a path whose plane is tilted at an angle of 23.5° to the plane of the equator.

In Copernicus' heliocentric system also the apparent looping of the planets can be easily explained. Let us take Mars, for example. This planet is first seen in the prolongation of the Earth's position (1 and 2). As the Earth, which revolves faster around the Sun, overtakes this planet, however, Mars appears to take on a retrograde motion in its orbit (in the interval between

---

6   Alexandre Koyré, *The Astronomical Revolution*, 27.

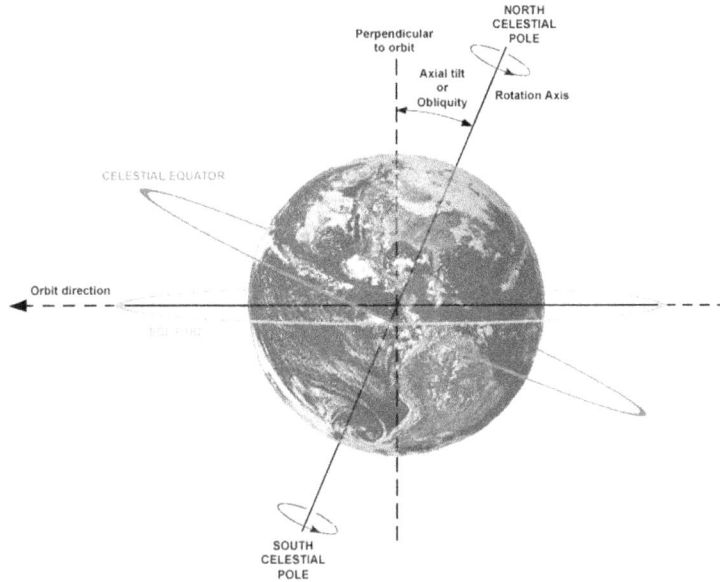

Figure 7. Axial tilt of the earth

Source: Wikipedia (http://en.wikipedia.org/wiki/Axial_tilt), accessed September 19, 2014.

3 and 4), which in a next phase is followed by a repair of the direct motion (in the interval between 5 and 6). This gives the observer from Earth the impression that Mars in a first move is brought to a standstill, and then appears to run back and finally is seen as moving forward again.

In both cases, erroneous perceptions are unraveled by reducing them to their *true causes*: the Earth's daily axial rotation and its annual revolution about the Sun. This causal explanation is far more convincing than the

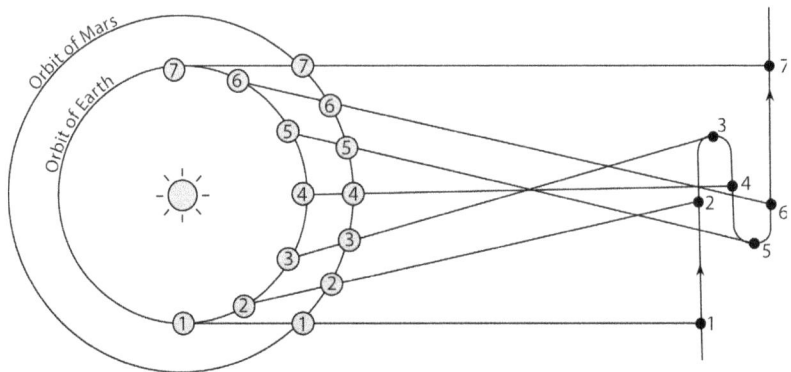

Figure 8. Looping of Mars

assumption that (i) the firmament revolves everyday at a fabulous speed, and (ii) that a supplementary eastward journey along the ecliptic makes the planets lag behind the firmament's daily westward rotation. The collapse of these two assumptions demonstrates the power of the Copernican revolution.

All this must remain pure speculation unless it is proven that the intuitively grasped simplification is also workable. Such a proof could only be given on the basis of geometrical calculations. Copernicus' ambitious project was to transpose Ptolemy's system of deferents and epicycles into the key of a heliocentric universe, and to demonstrate that this would result in a serious reduction of the number of epicycles. The inversion of middle points—Sun in the middle instead of the Earth—enabled Copernicus to discard five epicycles that in Ptolemy's geocentric universe were needed to explain the planets' apparent journeys along the ecliptic. They could be dropped because that route "appears to be nothing but the reflection of the annual revolution of the Earth about the Sun in the behavior of the planets."[7]

In spite of this success, Copernicus' program remained unfinished. One of the major obstacles to the completion of his work was that he continued to think in terms of rotating circles, while Kepler later would demonstrate that the planets' orbits are elliptical. So, a great many of the irregularities Copernicus perceived could not be resolved with the help of circular deferents and epicycles. In his effort to reduce some baffling composite motions to their regular components, he finally had recourse to an accretion of epicycles, in the best tradition of Ptolemy. His first draft of *De Revolutionibus Orbium Caelestium* (On the Revolutions of the Celestial Spheres), the *Commentariolus*, terminates with the declaration:

> Thus, Mercury moves on seven circles, Venus on five, Earth on three, and around it the Moon on four; and finally, Mars, Jupiter and Saturn each on five. Thirty-four circles in all suffice to explain the entire structure of the universe as well as the graceful motion of the planets.[8]

Books II to VI of *De Revolutionibus* abound with hypothetical constructions, thought experiments, and tentative calculations to determine whether some particular epicycle or accretion of epicycles was inconsistent with the overall model and had to be discarded. Among the commentators, this has raised the question as to the precise nature of Copernicus' new arrangement of the cosmic order. Did he regard his combinations of circles as purely

---

7   E. J. Dijksterhuis, *The Mechanization of the World Picture*, 292.

8   Alexandre Koyré, *The Astronomical Revolution*, 27.

hypothetical mathematical constructs that only exist in the mathematician's mind, or did he regard them as corresponding to the reality of the cosmos? This dilemma between hypothetical device and realistic description is not easy to resolve. Apparently, one ought to draw a clear distinction. Copernicus undoubtedly took it for granted that the planets revolved about the Sun as around their common focal center (the moon was something special). For the rest, he seemed not to have had enough evidence to account for all the precise mechanisms that regulated the intricate behavior of each of the planets, and there began the area of his hypothetical calculations.

History, however, apparently has its own surprises. When *De Revolutionibus* was published in 1543, it was accompanied by a preface that, to the amazement of many readers, "proclaimed the strictly hypothetical character of the new theory as well as all astronomical theories in general."[9] This preface was not signed, and although one might naturally have assumed that it expressed the author's opinion, it had, in fact, been written by Osiander, a German Lutheran theologian, who in private conversations and letters had repeatedly advised Copernicus to stress the *hypothetical* character of his theory. Copernicus had entrusted the printing of his work at Nuremberg to his disciple Rheticus. But since Rheticus had to leave for Leipzig unexpectedly, he delegated this task to Osiander, who took advantage of it and added his own preface to the work. The preface was a blow to Copernicus, for in his letter of dedication to Pope Paul III, he had underlined the astronomer's responsibility for giving a true picture of the cosmos.

The preface is entitled "To the Reader concerning the Hypotheses of this Work," and commences as follows:

> Since the novelty of the hypotheses of this work has already been widely reported, I have no doubt that some learned men have taken serious offence because the book declares that the Earth moves, and that the Sun is at rest in the centre of the Universe; these men undoubtedly believe that the liberal arts, established long ago upon a correct basis, should not be thrown into confusion. But if they are willing to examine the matter closely, they will find that the author of this work has done nothing blameworthy. For it is the duty of the astronomer to compose the history of the celestial motions through careful and skilful observations. Then turning to the causes of these motions or hypotheses about them, he must conceive and devise, since he cannot in any way attain to the true causes, such hypotheses as, being assumed, enable the motions to be calculated correctly from the principles of geometry, for the future as well as for the past. The present author has performed

---

9   Ibid., 36.

both these duties excellently. For these hypotheses need not to be true nor even probable; if they provide a calculus consistent with the observations that alone is sufficient.

The conclusion to the preface is even more eloquent:

So far as hypotheses are concerned, let no one expect anything certain from astronomy, which cannot furnish it; lest he accepts as the truth ideas conceived for another purpose, and depart from this study a greater fool than when he entered it. Farewell.[10]

Osiander had deliberately played down the cogency of Copernicus' arguments in favor of heliocentrism. His fear was that a dissemination of the Copernican theory had the potential to undermine the religious convictions of the Christian community. Heliocentrism, as a demonstrated fact and not merely a hypothetical construct, would have been a real shock to the average believer in the 16th century. This demonstrates the extent to which Christianity drew its worldview and its inspiration for worship not only from the biblical account of creation but also from Greco-Roman cosmology.

Tycho Brahe and Kepler

Copernicus still worked within a setting in which no considerable progress was made in the observation of celestial bodies. This situation changed with the advent of Tycho Brahe (1546–1601), who in his astronomical observatory in Uraniborg had kept a record of the altering positions of the planets. Brahe worked with a huge quadrant built into the wall of his study and centered over an open window through which he made observations. He had measured the longitudinal and latitudinal positions of about 1,000 stars, had registered comets, and had considerably corrected the Alfonsine Tables (and more recent ones), which contained the measurements of the changing positions of the planets over the years. His strength in observation, however, was not paralleled by daring theoretical insights. In a gesture of compromise, he held the view that the five planets revolved about the Sun, as about their provisional centre, whereas this centre orbited the immovable earth in the middle of the universe.

In 1600, a year before Brahe's death, Johannes Kepler (1571–1630) became Brahe's assistant. He regarded this appointment as a sign of divine

---

10   Ibid.

benevolence, as, given his natural propensity to engage in geometrical and even mystical speculations, he felt he was badly in need of greater familiarity with the results of empirical observation. On the other hand, as we will see, his speculative mind allowed him to ask the right questions in the elaboration of a celestial physics. Kepler

> devised the three laws that turned the system of Copernicus from a general description of the Sun and the planets into a precise, mathematical formula. First, Kepler showed that the orbit of a planet is only roughly circular: it is a broad ellipse in which the Sun is slightly off centre, at one focus. Second, a planet does not travel at constant speed; what is constant is the rate at which the line joining the planet to the Sun sweeps out the area lying between its orbit and the Sun. And third, the time that a particular planet takes for one orbit—its year—increases with its (average) distance from the Sun in a quite exact way.[11]

Mysterium Cosmographicum

In his youth, Kepler was uncertain as to whether he should become a pastor in the Lutheran church or an astronomer. In his first work, *Mysterium Cosmographicum* (1596), he employed *trinitarian* terms to evoke the majesty of the heliocentric cosmos. He compares the radiant Sun in the middle to the Father, and places the Son in the vault of the fixed stars, whereas the Holy Spirit fills the space in between with his divine energy.

In addition, Kepler set out to reconstruct the geometrical forms the Creator must have had in mind when creating the solar system. He did so in order to get an idea of the distances that separated the planetary spheres one from the other. He took the sphere of the Earth as his measure of all the other spheres—the outer spheres as well as the inner ones. He then began to fill up the space between these spheres with the regular solids that Plato held in great esteem: cube, tetrahedron, icosahedron, etc. This yielded the following result:

> The Earth is the measure for all other spheres. Circumscribe a Dodecahedron about it, then the surrounding sphere will be that of Mars; circumscribe a Tetrahedron about the sphere of Mars, then the surrounding sphere will be that of Jupiter; circumscribe a Cube about the sphere of Jupiter, then the surrounding sphere will be that of Saturn. Now place an Icosahedron within the sphere of the Earth, then the sphere

---

11   Jacob Bronowski, *The Ascent of Man* (London: British Broadcasting Corporation, 1973), 221.

which is inscribed will be that of Venus; place an Octahedron within the sphere of Venus, then the sphere which is inscribed is that of Mercury.[12]

Kepler submitted this result to Tycho Brahe, but he met with the latter's disapproval. Brahe found it an interesting thought experiment, but seriously

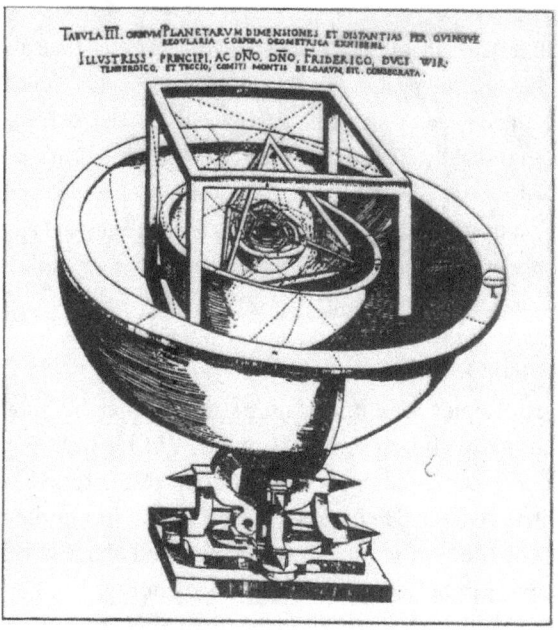

Figure 9. Kepler's Platonic solid model of the solar system

Source: Wikipedia (http://en.wikipedia.org/wiki/Johannes_Kepler), accessed September 19, 2014.

doubted the use of *a priori* constructions in astronomy because they threatened to overlook the data of empirical observation. Kepler would never forget Brahe's warning. Nevertheless, without his basic intuition about the calculability of the distances of the planets' orbits from the Sun, he would never have been able to formulate his Third Law. Before arriving at the Third Law, however, he had to abandon—through trial and error—two tenets he shared with his contemporaries: the spherical form of the planetary spheres, and the perfect uniformity of their rhythm of motion

---

12   Alexandre Koyré, *The Astronomical Revolution*, 146.

Astronomy with the Ellipse

Kepler laboriously discovered the elliptical shape of the planetary revolutions in various tentative steps. The first step, his study of the *eccentricity* of the orbit of Mars, would land him in a blind alley. In this study, he made use of Brahe's precious information about the changing positions of that planet in the course of the years. He set out to develop an eccentric model that would account for the observed positions of Mars in 1587, 1591, 1593, and 1595. The model he used was a "circle with divided eccentricity," in which (i) the Sun (E) is placed not in the center of the circle but at a lower point on the line of the apsides AD, and (ii) the radius FQ that carries the planet's deferent (and epicycle) around the circumference of the circle (in the direction of B) has its focus in the equalizing point Q. This was the model Ptolemy had used. Ptolemy had introduced the equalizing point (*punctum aequans*), also termed the "equant," to give the irregular motions of the planets a semblance of regularity.

Kepler modified Ptolemy's scheme in two respects. First, he placed the Sun where Ptolemy had placed the Earth, and, second, he abandoned Ptolemy's equidistant scheme in which the distance equalizing point-center (QC) exactly corresponded to the distance center-Sun (CE): he gave the Sun a lower place in the ovoid whole (thus putting the Sun more or less in one of the foci of an ellipse, although at that moment, he did not yet realize the ellipse was the correct model to describe Mars's orbit).

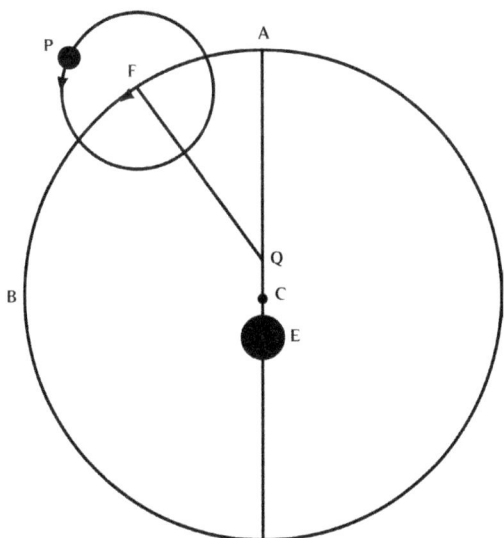

Figure 10. Equalizing point and equidistant hypothesis

Looking at the result he found out that he was able to account for the longitudinal positions of Tycho Brahe's observations, which all did fit in the new construction (with errors not exceeding 2' 12'' which could be attributed to minor inaccuracies in Brahe's observations). Yet, he did not succeed in accounting for the latitudinal positions. So, the working hypothesis could not be trusted, for longitudinal positions only show the *direction* in which a planet can be seen from the Sun at a particular moment, but do not tell anything about the planet's *distance* from the Sun, which can only be known by determining latitudinal positions as well.

Discouraged by this result, Kepler had recourse again to the equidistant scheme. Had Ptolemy worked with it, he wondered, because it was so symmetrical, or because he had found it really corresponded to observed facts? In this return to Ptolemy he experienced the surprise of his life. The equidistance hypothesis enabled him again "to account with sufficient accuracy for the heliocentric longitude of the 4 observed positions."[13] But trying to determine also the respective latitudes of the planet's positions, he obtained quite satisfactory agreement between calculated and observed positions for the points close to the apsides H and I (namely at F and D) and close to positions 90° away from them (namely at G and E), "but at intermediate positions, corresponding to displacements of 45° or 135°, the discrepancy between calculation and observed fact amounted to 8'."[14] In chapter 19 of the *Astronomia Nova*, he expresses his gratitude for what he learned from this discrepancy. He wrote:

> For us, to whom God's goodness has given in Tycho Brahe a most careful observer, from whose observations the error of 8' in the Ptolemaic calculations is revealed, it is fitting to recognize with a grateful heart this good gift of God and make use of it. Let us, therefore, labour finally to trace out the true nature of celestial motions, basing ourselves on the evidence for the incorrectness of the suppositions made.[15]

As we know today, the discrepancy between Kepler's construct and observed facts flows from the *elliptical shape* of Mars's orbit. Yet, at that moment Kepler only realized that there was something wrong with the construct, without being able to get at the root cause of the error. So, he interrupted his study of the orbit of Mars awhile, and turned his attention to the Earth's revolution about the Sun. This examination had to be done first,

---

13   Ibid., 402.

14   Ibid., 178.

15   E. J. Dijksterhuis, *The Mechanization of the World Picture*, 307.

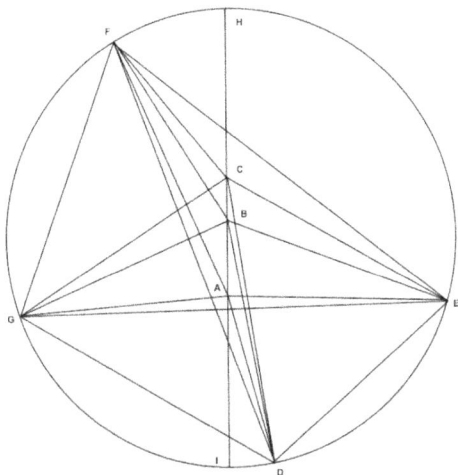

Figure 11. Scheme resulting in an error of 8′

for as long as one had no certainty about the exact shape of the Earth's orbit about the Sun, errors in this matter would be projected onto the apparent behavior of the other planets.

In his examination of the Earth's orbit, Kepler used an original method. He set out to calculate how the Earth's orbit would look when observed from a fixed point on the orbit of Mars. Since Mars takes 687 Earth days to orbit the Sun once (whereas it takes the Earth only 365 days to do the same), Kepler was able to determine the positions of Mars that were exactly 687 Earth days apart. With this approach, Mars would be found to be in the same place in its orbit after each circulation about the Sun, whereas the Earth, on the 687th day, would be in a different location. By analyzing these different locations Kepler succeeded in calculating the Earth's positions at its farthest and shortest distance from the Sun. He then jumped to the conclusion that (a) a planet's path must be elliptical, and (b) that there must be an exact ratio between a planet's momentary distance from the Sun and its acquired speed. If the planet's orbit is farthest away from the Sun (i.e., in its *aphelion*), the planet moves the slowest; if, on the contrary, it is nearest to the Sun (i.e., in its *perihelion*), it moves the fastest.

This insight will lead to the formulation of his second law, the law of areas. It takes the same amount of time to go from A to B as it does to go from C to D. The reason for this is quite simple: the respective regions have the same area. This means that between C and D the planet must move faster about the Sun than between A and B. The shorter the radius vector, the nearer the planet is to the Sun, and thus the faster it moves. Conversely, the larger the radius vector and the farther away the planet is from the Sun, the slower it goes.

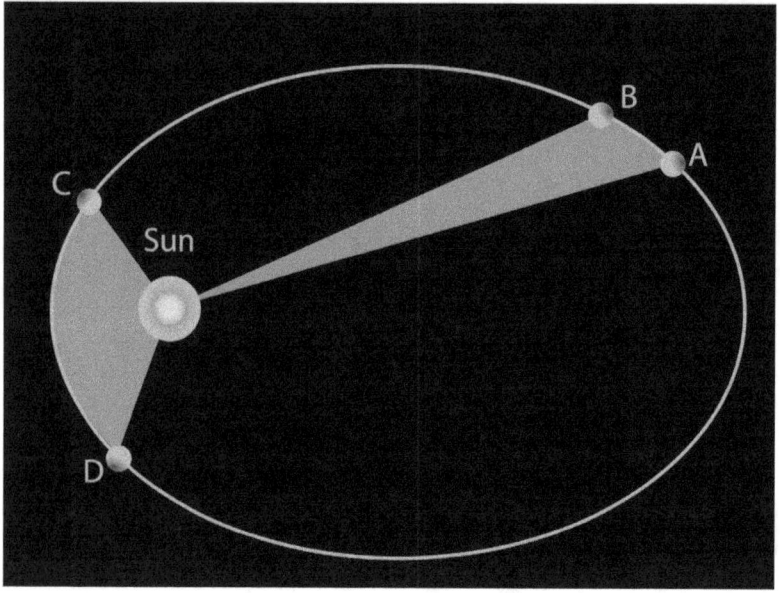

Figure 12. Kepler's law of areas

Source: Windows to the Universe (http://www.windows2universe.org/ the_universe/uts/kepler2. html), accessed September 19, 2014.

In his last work, *Harmonice Mundi* (1619)—The harmony of the world—Kepler revisits Plato's music of the spheres. He sought to retrieve this theory after the collapse of the two cosmological dogmas, viz., the circularity of the planetary motions and their uniform velocities. For this reason, the music of the spheres can never again be what it had been. From now on, this music will be determined not only by precise distances but also by kinetic laws. For Kepler, each planet emits a particular melody made up of a few notes and whose sound depends on its changing speed of revolution. True, just as in Plato's model, the more distant planets like Saturn and Jupiter, with a longer periodical time of revolution give off a deeper melody than the nearer planets with their shorter year of revolution. Yet, what is typical of all the planets is that, due to the changing speeds on their elliptical orbits, their melodies are no longer just made up of regular octaves, but of rhythmic modulations depending on their phase of acceleration or retardation. Transposed in whole and half-tones, the melody of Saturn is a deep "sol, la, ti, la, sol," that of Jupiter is, in a higher octave, "ti, do, re, do, ti." Mars sings in a still higher octave "fa, sol, la, ti flat, do, ti flat, la, sol, fa," and the Earth, in a still higher octave, "sol, la, sol." Venus, in turn, whose ellipse is not very elongated, sings practically a still higher monotone "mi, mi, mi."[16] Kepler was eager to transpose the music of the spheres into the key of the new cosmology. Worship of God was for him linked with the true knowledge of nature. His work ends with a prayer: "Please, God, grant that my demonstrations may serve Thy Glory and never be deleterious to the salvation of souls."[17]

Galileo's Condemnation

Kepler has three astronomical laws to his name. Galileo (1564–1642), however, acquired his place in the history of science through his experiments with the free fall, which laid the foundation of modern mechanics. His name is also associated with the ignominious trial staged against him because of his proofs in favor of the Copernican revolution. He was mainly active on two fronts: the observation of the sky and the invalidation of Aristotelian physics. The results he obtained in these domains convinced him that Copernicus' theory was correct: the Sun and not the Earth is the center of the planetary system. This conclusion brought him into conflict with the Church, whose theologians were strongly dependent on Aristotle's picture of the world.

In 1589 Galileo was appointed to the chair of mathematics and physics at the University of Padua in the territory of the Doge of Venice. In 1610, a year after the publication of Kepler's *Astronomia nova*, he published a book: *Sidereus nuncius,* or the "Starry Messenger," which stirred a nationwide commotion. In this book, he reported astronomical data detected with his telescope. This instrument was a novelty, the improved version of a spyglass of Dutch origin that consisted of a convex and concave lens in a tube with the effect of magnifying objects three or four times. When Galileo thought his telescope was sufficiently perfected—by then it could magnify objects eight to ten times—he invited some Venetian citizens and senators to watch, with the help of

---

16   Alexandre Koyré, *The Astronomical Revolution*, 337.

17   Johannes Kepler, *Harmonice mundi libri V* (1619), Gesammelte Werke, ed. Van Dijck, W. Casper, vol. VI (Munich, 1943), 243 (translation mine).

this instrument, ships approaching the city's harbor. They did so from the top of the Campanile, and were able to identify ships that were two or more hours' sail away.

The telescope enabled him to detect new astronomical data that he brought out in his book. He discovered that the Milky Way, which until then was perceived as a nebulous conglomerate, was formed by distinct groups of stars. He also observed Sunspots and Moon craters, which was an occasion for him to shed further doubt on Aristotelian physics: indeed, how could a celestial body with vast protuberances and deep chasms be made of quintessence? His most sensational discovery, at that time, however, was the evidence of four large moons (now called the Galilean satellites) that orbited Jupiter. From this discovery, he concluded that, if moons could orbit Jupiter, why could the Earth—with its own moon rotating about it—not orbit the Sun? After the publication of his book, Galileo further discovered that not only the Moon but Venus as well had its proper phases, and that Saturn appeared to be divided into three regions, "the first observation of what Huygens (1629–1695) was to recognize as a ring about that planet."[18]

Galileo's observations met with skepticism. His telescope was regarded as a strange instrument, brought into circulation by vendors from the North. In addition, he was not a professional astronomer, which to a certain extent is true, for in his writings one will look in vain for new data concerning the longitudinal and latitudinal positions of stars and planets. His specialty was rather the study of the kinetic laws of falling bodies on Earth. It is in this field that he performed successful experiments *indirectly* demonstrating that the twofold rotation of the Earth had no impact on the validity of these laws.

Galileo's most famous experiment is that of the fall movement staged on an inclined plane. "He rolled balls of different weights down a smooth slope. The situation is similar to that of heavy bodies falling vertically, but it is easier to observe because the speeds are smaller. Galileo's measurements indicated that each body increased its speed at the same rate, no matter what its weight. For example, if you let go off a ball on a slope that drops by one meter for every ten meters you go along, the ball will be travelling down the slope at a speed of about one meter per second after one second, two meters per second after two seconds, and so on, however heavy the ball. Of course, a lead weight would fall faster than a feather, but that is only because a feather is

---

18   E. J. Dijksterhuis, *The Mechanization of the World Picture*, 381.

slowed down by air resistance. If one drops two bodies that don't encounter much air resistance, such as two different lead weights, they fall at the same rate."[19] This example reveals the precision of Galileo's mechanics.

Was it because his kinetic theory contained ideas concerning the mobility of the Earth that Galileo ran into trouble with the Holy Office of the Catholic Church? In any case, dark clouds started to gather as soon as he settled in Florence, where he enjoyed the protection of the influential family de Medici. From 1613 onward, reports were filed about his teaching and publications, and in 1616, like the thunderbolts of Jupiter, came the posthumous condemnation of Copernicus, which indirectly was also aimed at him: "Propositions to be forbidden: that the Sun is immovable at the centre of the universe; that the Earth is not at the centre of the heavens, and is not immovable, but moves by a double motion."[20] These propositions, it was said, were contrary to the Catholic faith. In Ps. 104:5, one reads, "Thou [God] did fix the earth on its foundation, so that it never can be shaken," and elsewhere Scripture tells us that Joshua commanded the Sun to stand still [see Josh. 10:12–14], and not the Earth.

The Holy Office of the Inquisition was reorganized in 1542, in the period of the Counter-Reformation; it was the supreme ecclesiastical organ for scrutinizing heretical opinions. The German Kepler, who lived in a Protestant country, was immune to such an examination (though in his homeland, his mother had been persecuted for witchcraft), but Galileo, who was a Catholic Italian, could be summoned for trial at any moment. Indeed, Copernicus' posthumous condemnation not only meant that his works were placed on the list of forbidden books (where they remained until 1835) but also that writings propagating his doctrine were likewise prohibited.

Given these gloomy circumstances, Galileo thought it would be wise to consult Cardinal Bellarmine, an influential Jesuit theologian at the papal court. From Bellarmine, he got the advice to go on with his cosmological studies, but to prudently present them as the outcome of conjectural hypotheses. Bellarmine apparently knew of Aquinas' view, which Osiander had also followed: astronomy could only yield uncertain, hypothetical results. As a science, it lacked the certainty that was to be associated with Sacred Theology. In actual fact, Bellarmine's

---

19   Stephen Hawking, *A Brief History of Time* (London: Bantam Press, 1989), 15–16.

20   Jacob Bronowski, *The Ascent of Man*, 207.

appeal to scholastic authority was intended as a diplomatic cloak to safeguard Galileo's research. His Jesuitical suggestion implied encouragement as well as caution.

Realizing his precarious situation, Galileo attempted to get papal protection. He saw a window of opportunity in 1623 with the election as pope of Maffeo Barberini, a great intellectual. With this pope, who had adopted the name of Urbanus VIII, Galileo had a conversation that was to become fatal to him. He informed the pope that he was planning to write a book on the two chief world systems from a purely scientific standpoint. Counting on the approval of the pope, he even ventured to say that "in discussions of physical problems one ought to begin not from the authority of scriptural passages, but from experiences and necessary demonstrations [...] Nor is God any less excellently revealed in Nature's actions than in the sacred statements of the Bible." The pope, however, got furious and replied "that there could be no ultimate test of God's design," and insisted that Galileo's book support this principle, "for it would be an extravagant boldness for anyone to go about to limit and to confine the Divine power and wisdom to some one particular conjecture of his own."[21]

In 1632, Galileo published his *Dialogue Concerning the Two Chief World Systems.* It was written in Italian and presented as a series of discussions, over a span of four days, among two philosophers and a lay man. The philosopher Salviati argued for the Copernican position and also presented some of Galileo's views. The intelligent lay man Sagredo took a stand of neutrality, and the philosopher Simplicius presented the "official" point of view, that is: Aristotelism. Simplicius was named after a sixth-century commentator on Aristotle, but in Italian this name also referred to one who was "simple-minded." It is into his mouth that Galileo put the pope's words: "There is no ultimate test of God's design." The book had been submitted to local ecclesiastical censorship, and permission had been given for the work to be printed. When it came out, it was a best-seller. The pope was so outraged that he ordered that the book be taken out of print, that all the copies that had been sold were to be bought back, and that Galileo be summoned to Rome to account for his impertinence. The ominous order reads as follows: "His Holiness charges the Inquisitor at Florence to inform Galileo, in the name of the Holy Office, that he has to appear as soon as possible in the course of the month of October before the Commissary-General

---

21   Ibid., 209.

of the Holy Office."[22] The pope himself had delivered Galileo into the hands of the Inquisition.

At his trial, Galileo, nearly seventy years old, was severely interrogated and compelled to retract the statements he had made in favor of the Copernican system. He was even shown the instruments of torture, with the intimation that they might be used on him. When one reads the acts of the trial in retrospect, it is shocking to see that not a word was spent on properly discussing the subject-matter of the two chief world systems. The sole issue to be treated was the question as to whether Galileo had abided by the prohibitions issued in 1616. It looked as if his condemnation was a predetermined outcome of the trial.

In the course of his interrogation Galileo referred to the meeting he had with Cardinal Bellarmine in February 1616. He also showed the inquisitors the certificate Bellarmine had sent him on 26 May 1616 attesting, first, that he, Galileo, "had never abjured [i.e., had never had to abjure] anything," and second, that he was prohibited to "hold or to defend the said opinion"[23] (which indirectly implied that the Copernican system could be taken and used as a conjectural supposition). For Galileo, this certificate was crucial, for in conformity with it he had written his book in the form of a dialogue in which the pros and cons of the working hypothesis were debated. So, personally he had "not held or defended" the Copernican doctrine. The inquisitors retorted that from the tone of the book one could deduce that the author sympathized with Copernicus; they resented Galileo's maneuver for bypassing the prohibition of 1616, and required him to formally abjure the heretical doctrine. They insisted that he sign the following retraction:

> I Galileo Galilei, son of the late Vincenzo Galilei, Florentine, and aged seventy years, arraigned personally before this tribunal, and kneeling before you, most Eminent and Reverend Lord Cardinals, Inquisitors General against heretical depravity throughout the whole Christian Republic, having before my eyes, and touching with my hands the Holy Gospels swear that I have always believed, do now believe, and by God's help will for the future believe, all that is held, preached and taught by the Holy Catholic and Apostolic Roman Church. But whereas—after an injunction had been judicially intimated to me by this Holy Office, to the effect that I must altogether abandon the false opinion that the Sun is the centre of the world and immovable, and that the Earth is not the

---

22   Ibid.

23   Ibid.

centre of the world, and moves, and that I must not hold, defend, or teach in any way whatsoever, either orally or in writing, the said doctrine, and after it had been notified to me that the said doctrine was contrary to Holy Scripture—I wrote and printed a book in which I discuss this doctrine already condemned, and adduce arguments of great cogency in its favor, without presenting any solution of these; and for this cause I have been pronounced by the Holy Office to be vehemently suspected of heresy, that is to say, of having held and believed that the Sun is the centre of the world and immovable, and that the Earth is not the centre and moves. Therefore, desiring to remove from the minds of your Eminences, and of all faithful Christians, this strong suspicion reasonably conceived against me, with sincere heart and unfeigned faith I abjure, curse, and detest the aforesaid errors and heresies [...] I, the said Galileo Galilei, have abjured, sworn, promised, and bound myself as above, and in witness of the truth therefore I have with my own hand subscribed the present document of my abjuration, and recited it word for word at Rome, in the Convent of Minerva, this twenty-second day of June, 1633.[24]

The trial was ignominious. One of the inquisitors accused Galileo of not having complied with the instruction addressed to him in which it was clearly stated that "he must not hold or defend the said opinion, *or teach it in any way whatsoever, either orally or in writing*" (thus not even by way of discussing it as a *hypothesis*). The instruction to which the inquisitor alluded was not shown to Galileo; in its view, the Holy Office was not obliged to give evidence of its accusations. Galileo replied that he knew nothing of that instruction, and referred again to Cardinal Bellarmine's certificate, which was less restrictive. The instruction in question is in the secret archives of the Vatican, but its authenticity is quite uncertain. Had it been drafted before or after the trial? Some suspect that it was a forgery or, "at the most charitable, a draft for some suggested meeting which was rejected. It is not signed by Cardinal Bellarmine. It is not signed by the witnesses. It is not signed by the notary. It is not signed by Galileo to show that he received it."[25] It took almost 360 years before the church deemed it necessary to rehabilitate Galileo. It was only in October 1992, twenty-three years after the first man landed on the Moon (1969) and fifteen years after the take-off of Voyager 2 (1977), programmed to reach the planet Neptune by 1989 after having covered 4.43 billion miles of interplanetary space,

24   Ibid., 216.
25   Ibid., 214.

that Pope John-Paul II publicly deplored the church's condemnation of Galileo and of the Copernican world system.

Galileo was confined to his villa near Florence for the rest of his life. There he finished his great work on physics, *Discorsi*, in which he went on to elaborate his argument that one set of physical laws must suffice to explain the kinetic phenomena on Earth and in the heavens. Yet, he only succeeded in laying down the groundwork for this all-encompassing view. Although Galileo was successful in undermining Aristotelian physics, his laws of motion still lacked the comprehensiveness that would allow them to predict the behavior of the planets. One will have to wait for Newton to formulate the universal laws of motion that provide a satisfactory explanation for the whole of physical reality.

Galileo's specialty was the study of the acceleration of motion. In addition, he was aware of the strange equivalence that exists between the state of rest and the state of unaltered straight line motion. Indeed, the reason why we do not feel the Earth's daily axial rotation is that this motion is so uniform and always in the same direction. In *Dialogue Concerning the Two Chief World Systems* he illustrated this equivalence with the following example:

> Shut yourself up with some friends in the main cabin below decks on some large ship, and have with you there some flies, butterflies, and other small flying animals. Have a large bowl of water with some fish in it; hang up a bottle that empties drop by drop into a wide vessel beneath it. With the ship standing still, observe carefully how the little animals fly with equal speed to all sides of the cabin. The fish swim indifferently in all directions; the drops fall into the vessel beneath.[...] When you have observed all these things carefully [...] have the ship proceed with any speed you like, *as long as the motion is uniform and not fluctuating this way and that.* You will discover not the least change in all the effects named, nor could you tell from any of them whether the ship was moving or standing still [...]The droplets will fall as before into the vessel beneath without dropping towards the stern, although while the drops are in the air the ship runs many spans. The fish in their water will swim towards the front of their bowl with no more effort than toward the back, and will go with equal ease to bait placed anywhere around the edges of the bowl. Finally the butterflies and flies will continue their flights indifferently toward every side, nor will it ever happen that they are concentrated toward the

stern, as if tired out from keeping up with the course of the ship, from which they will have been separated during long intervals by keeping themselves in the air.[26]

For Galileo, "uniform straight-line motion is physically completely indistinguishable from the state of rest (i.e., of absence of motion)."[27]

---

26   Galileo Galilei, *Dialogue Concerning the Two Chief World Systems*, trans. S. Drake (Berkeley: University of California, 1953), 186–97 (italics mine).

27   Roger Penrose, *The Emperor's New Mind: Concerning Computers, Minds and the Laws of Physics* (Oxford: Oxford University Press, 1989), 163.

CHAPTER 7

Newton

Isaac Newton (1642–1727) was born in the same year Galileo died. His name is associated with the birth of the mechanistic world picture. Newton's major achievement, the *universal* law of gravitation, would have been unthinkable, though, without Kepler's laws for planetary motions and Galileo's law for falling bodies, which Newton unified to form a higher synthesis. Yet, he would never have succeeded in this enterprise had he not devised the instrument on which this synthesis rests, "the formulation of the differential or 'infinitesimal' calculus," which is concerned with the study of the rates at which quantities change.[1] Here, one really felt the pulse of the new epoch: the rise of engineering.

For the new generation of scientists, all things in nature could be measured and determined.

> Take that body of ideas that came together in the seventeenth and eighteenth centuries under the heading of "classical science" or "Newtonianism." They pictured a world in which every event was determined by initial conditions that were, at least in principle, determinable with precision. It was a world in which chance played no part, in which all the pieces came together like cogs in a cosmic machine.[2]

Newton's laws of motion were hallowed in England and throughout the whole of Europe as the breakthrough of scientific rationality; their

---

1   See Ilya Prigogine and Isabelle Stengers, *Order out of Chaos: Man's New Dialogue with Nature* (London: Fontana Paperbacks, 1986), 57.

2   Alvin Toffler, *Foreword* in *Order out of Chaos* by Ilya Prigogine and Isabelle Stengers, xiii.

author was celebrated as the New Moses, who had been shown the 'tablets of the law'. These tablets contained, with mathematical precision, all the rules with which nature had to comply. On the occasion of Newton's death in 1727, the poet Alexander Pope composed the following epitaph:

> Nature and Nature's law lay hid in night:
> God said, let Newton be! and all was light.

Another poet enthusiastically wrote:

> Nature compelled, his piercing Mind obeys,
> and gladly shows him all her secret Ways;
> 'Gainst Mathematicks she has no Defence,
> and yields t' experimental Consequence.[3]

Newton, however, was, strictly speaking, not a follower of "Newtonianism": he was not a rationalist in the sense that he would have seen a conflict between reason and belief in God. He firmly believed in the dominion of God, the Creator, over the laws of nature. For him, God is the Lord and Ruler whose omnipresence makes the laws of nature work. His mindset is miles apart from that of the Frenchman Laplace (1749–1827), who was "convinced that science had *proved* that nature is transparent." When asked by Napoléon what God's place was in his world system, he laconically replied: "Majesty, I do not stand in need of this hypothesis."[4]

Isaac Newton was twenty-six years old when he became a Lucasian professor of mathematics at Trinity College in Cambridge. His chair was apparently a sponsored chair, a peculiarity that sheds some light on the overall climate of that century. In the 17th century, research into the new positive sciences was mostly done outside the classical universities, which until in the Renaissance, confined themselves to the study of the humanities, philosophy, medicine, law, theoretical mathematics and cosmology. Typical of that century was the rise of the Academies and Royal Societies that sponsored research and offered a seat for scientists to discuss their new discoveries. The Royal Academy of Sciences in Paris, e.g.,

> sponsored the measurement of the length of one minute of arc on the Earth's surface, thus determining the size of the Earth with an accuracy far beyond any

---

3   Ilya Prigogine and Isabelle Stengers, *Order out of Chaos*, 27.
4   Ibid., 52.

earlier measurement. An expedition to South America helped to determine the distance of Mars from the Earth and by indirection the dimensions of the solar system. Projects of similar scope were beyond the means of individual scientists and the *Académie des Sciences* put the entire scientific community under obligation by conducting them.[5]

Such sponsorship put serious strains upon the scientists, not only in Paris, but also at the Royal Societies in London and Berlin (where Leibniz was based), and created a climate of (international) scientific competition. No wonder then that a gifted mathematician like Isaac Newton, already from his youth, applied himself to various branches of learning, such as the infinitesimal calculus, the phenomenon of gravitation, and the refraction of light—when light passes through a prism it breaks up into the colors of the spectrum, and these colors can be rejoined to produce a ray of white light when they are passed through a second prism.

In 1672, four years after his appointment as Lucasian professor of mathematics, Newton was elected into the Royal Society of London, of which he became the president after the publication of his *Principia Mathematica* in 1686. He was not a pleasant man when engaged in disputes; he distrusted his colleagues for fear they might fraudulently publish the provisional results of his investigations, which he kept secret till they were entirely confirmed. His obsession with rivalry brought him into conflict with the German philosopher Gottfried Wilhelm Leibniz. A quarrel arose over who was the first to "develop a branch of mathematics, called calculus, which underlies most of modern physics."[6] Newton hardly accepted criticism. His work *Opticks* was already substantially finished in 1666, but he published it only in 1704, a year after the death of his opponent, Robert Hooke. Although he firmly believed in the Creator God, he repudiated the doctrine of the Trinity, and—something not to be expected from a rational scientist—was engaged in alchemy.

The Laws of Motion

The story goes that Newton began to be intrigued by gravitation when being hit by an apple that fell from a tree under which he was sitting in his garden. This event may have stirred his imagination, but he had still a long way to go before he was able to discover the "incredible notion

---

5   Richard Westfall, *The Construction of Modern Science: Mechanisms and Mechanics* (Cambridge: Cambridge University Press, 1984), 111.

6   Stephen Hawking, *A Brief History of Time* (London: Bantam Press, 1989), 192.

of Universal Gravitation—the idea that all massive bodies continuously attracted all other bodies according to a precise mathematical law."[7] The *Principia Mathematica*, in which this universal law figures, commences, in fact, with a systematic study of the general laws of motion, critically examined against the background of the findings of Galileo and Descartes.

Antecedents: Galileo and Descartes

Newton had acquainted himself with the geometry of René Descartes (1596–1650) when he was a student:

> He largely taught himself mathematics through extensive reading of recent publications, most notably of the second edition of van Schooten's Latin translation, with added commentary, of Descartes's *Géométrie*. Within an incredible short period, less than two years, Newton mastered the subject of mathematics, progressing from a beginning student of university mathematics to being, *de facto*, the leading mathematician of the world.[8]

In his *Géométrie* Descartes had integrated and systematized the two major achievements of Galileo: the description, in mathematical terms, of both uniform motion and uniformly accelerating motion. It is also from Galileo that he learned (i) the importance of the notion of inertia (So long as there is no force that is going to change a body's state of rest or uniform, straight line motion, this state of affairs will persist) and (ii) the principle of the conservation of energy: in order to make a body return to the height from which it was dropped, one would need the same amount of energy that was needed to make it fall.

Galileo contented himself with proposing mathematical solutions to physical problems. Yet, Descartes, in addition, also embarked on the elaboration of a mechanical philosophy. In this light he separated the domain of human intelligence (*res cogitans*) from that of matter (*res extensa*): this was, for him, a prerequisite for understanding the working of the nature: "The world is a machine, composed of inert bodies, moved by physical necessity, indifferent to the existence of thinking beings."[9] For Descartes, the material world is characterized by extension. Matter is coextensive with space,

---

7   Rob Lliffe, *Newton: A Very Short Introduction* (Oxford: Oxford University Press, 2007), Preface.

8   Bernard Cohen and George Smith, *Introduction* in ID, *The Cambridge Companion to Newton* (Cambridge: Cambridge University, 2002), 10.

9   Richard Westfall, *The Construction of Modern Science*, 33.

which implies that the whole of space is filled with particles of matter. Particles of matter are by definition inert stuff pruned of conscious action, but they are capable of interacting with one another and of changing their place in the universe in accordance with the mechanical laws of motion.

Descartes' theory of motion is best understood when read against the background of nascent deism. "In the 17th century, everyone agreed that the origin of motion lay with God. In the beginning, He created matter and set it in motion." Once set in motion, matter obeys the principle of inertia. This principle practically flows from God's divine immobility: "Every particular part of matter continues always in the same state as long as the encounter of other parts of matter does not force it to change"[10] (First Law). Yet, because of the world's difference from God the material particles can have their primordial state changed in conformity with the "law of change": motion is added to a body through subtraction from another body: "When one body pushes another, it cannot impart any motion to the other, unless it loses at the same time as much of its own; nor take away from the other unless its own increases that much"[11] (Second Law). Through this mechanism the total quantity of motion is conserved; it cannot grow or diminish. "The supreme law of the world is the law of constancy, or conservation. [...] Things started to move at the same time the world was created; and, this being so, it follows therefrom that this motion will never cease, but only pass from subject to subject."[12]

Descartes' influence on the intellectual life of 17th century Europe was enormous. Cartesianism replaced Aristotelian scholasticism as the only intellectually respectable option. Even its cosmological speculations about the existence of a vortex made up of tiny particles wheeling about the Sun (and other Suns in the universe) would seduce scientists like Christian Huygens (1629–1695) and Gottfried Leibniz (1646–1717).

Descartes introduced the vortex theory to account for the planets' common revolution about the Sun, at regular distances from one another. To a certain extent, he succeeded in making his theory plausible. The basic presupposition was, again, that the whole space of the universe was entirely filled with matter because of the equation of extension with matter. So, in order to theorize a circular motion about the Sun, he had to posit (i) that the material particles involved were thinner than matter elsewhere, and (ii) that these tiny particles were able to whirl about the Sun by vacating the

---

10   Alexandre Koyré, *Newtonian Studies* (London: Chapman & Hall, 1965), 72.

11   Ibid.

12   Ibid., 71.

space into which they moved. This displacement created a closed circuit of moving matter, like the rim of a wheel turning on its axis, in which the planets were "swimming." In order to further explain the distances between the orbiting planets, Descartes had recourse to the centrifugal force that supposedly makes the planets recede, by degrees, from the center.

> As the whole vortex whirls about its axis, every particle in it endeavors to recede from the center, but in a plenum [i.e., in an entirely filled space] one particle can move away from the center only if another moves toward it. Like every other body, each planet tends to recede from the center, but at some distance from the center its tendency to recede is simply balanced by the swiftly moving matter of the vortex beyond it. An orbit is established by the dynamic balance between the centrifugal tendency of a planet and the counter-pressure arising from the other matter composing the vortex.[13]

The vortex theory was designed to replace the Aristotelian crystalline spheres with a model that suited the new mechanical picture of the universe. The machine-like process of centrifugal tendency and counter-pressure "explained" why the planets, at varying distances, all orbit in the same direction and in the same plane about the Sun, but it also contained some dubious elements reminiscent of the old cosmology, such as the assumption that the particles in the gyrating vortex were of a subtler quality than ordinary matter. In spite of its highly speculative character, the theory did not meet with much criticism because it was framed in the mechanistic pattern that was widely accepted in those days. For Newton, however, the vortex theory lacked scientific rigor since Descartes never developed a mathematical explanation for it.

Newton's Three Laws of Motion

Newton formulated his laws of motion in the *Principia Mathematica*. The first law, which is reminiscent of Descartes' principle of inertia, states that "every body continues in its state of rest, or of uniform motion in a right line, unless it is compelled to change that state by forces impressed upon it."[14] The formulation of this law is very precise especially in the Latin version, which uses the verb *perseverare in*, whereas the English translation says "continues in." Thus, a body perseveres in its state unless it is

---

13   Richard Westfall, *The Construction of Modern Science*, 35.

14   Isaac Newton, *Principia*, Vol. I: *The Motion of Bodies*, ed. Florian Cajori (Berkeley and Los Angeles: University of California Press, 1966), 13.

compelled by an external force to change that state. This state is either a point at rest or a point travelling with a uniform velocity in a right line. The second law, the law of change, states: "the change of motion is proportional to the motive force impressed; and is made in the direction of the right line in which that force is impressed."[15] This law allows one to calculate the acceleration brought about by the motive force ("If any force generates a motion, a double force will generate double the motion, a triple force triple the motion, whether that force be impressed altogether and at once, or gradually and successively.")[16] It also specifies the direction the acceleration is going to take.

With the formulation of his second law Newton considerably differs from Descartes. When dealing with the change of motion Descartes' main concern was to demonstrate that in spite of a transfer of motive force the total quantity of motion was conserved. If body A is to change the motion of body B, A will lose the exact quantity of motion that it transfers to the accelerating body B. For our discussion it is important to know what Descartes understood by "quantity of motion." Richard Westfall makes it clear that "by quantity of motion Descartes meant the product of that body's size and its velocity."[17] Newton apparently knew of this definition and corrected it.

First of all, Newton specified that a body's size alone cannot give us a precise idea of the motive force impressed: what matters is the quantity of *matter* the body in question contains. To measure that quantity of matter, one ought to take into account *the density* (a notion which is totally absent in Descartes' approach) and the bulk of that body:

> The quantity of matter is the measure of the same, arising from its density and bulk conjointly. Thus air of a double density, in a double space, is quadruple in quantity; in a triple space, sextuple in quantity. The same thing is to be understood of [...] all bodies that are by any causes whatever differently condensed [...] It is this quantity that I mean hereafter everywhere under the name of body or mass.[18]

Newton adds that "mass" or "quantity of matter" must not be confused with weight, although a body's mass is indirectly known by its weight. It is only after this specification that he sets out to define the quantity

---

15   Ibid.

16   Ibid.

17   Richard Westfall, *The Construction of Modern Science*, 121.

18   Isaac Newton, *Principia*, Vol. I: *The Motion of Bodies*, 1.

of motion that is going to be transmitted. This quantity or 'motive force impressed' is given by the product of the body's *mass* and its velocity. The more massive body A and the faster it moves, the more it will change (accelerate) the motion of body B with which it collides.

Newton, second, pointed out that velocity and acceleration are vector quantities, which are characterized by magnitude and direction: so, "the velocity's direction has to be taken account of as well as its magnitude."[19] This focus on the direction comes to the fore in Newton's second law: "The change of motion [i.e., the acceleration] is made in the direction of the right line in which that force is impressed."[20] On the basis of this definition, precise calculations can be made of the rate and the direction of the acceleration. In his formulation of the "law of change," Descartes seemed to have ignored the respective *vector* quantities of velocity and force. Thus, he was not able to determine the precise manner in which an impinging force A with a particular mass, velocity, and angle of collision was to catapult body B into a particular direction.

This double correction reveals Newton's superiority over Descartes. Descartes had only been able to declare that, "if a body in motion meets another stronger than itself, it loses none of its motion; and if it meets a weaker one which it can move, it loses as much of its motion as it gives to the other."[21] This declaration only explains the conservation of the total quantity of motion in the universe but fails to account for the exact rate of the catapulted acceleration that is going to take place.

Newton added a third law to the two laws of motion, the law of action and reaction: "To every action there is always opposed an equal reaction; or, the mutual actions of two bodies upon each other are always equal and directed to contrary parts."[22] This law says that the action body A exerts on body B is accompanied by an equal and opposite reaction which body B exerts on body A. So, the two bodies' forces exert a mutual and equal force upon each other. Apparently, Newton needed this law to make his picture of the world coherent, just as Descartes needed the theory of the vortex to explain how particles can whirl around through a space entirely filled with matter. Newton abhorred Descartes' notion of compact space. For him, the universe is sparsely populated with particles or mass points that move in the void of empty space. If this is the case, one has to

19   Roger Penrose, *The Emperor's New Mind* (Oxford: Oxford University Press, 1989), 162.

20   Isaac Newton, *Principia*, Vol. I: *The Motion of Bodies*, 13.

21   Richard Westfall, *The Construction of Modern Science*, 122.

22   Isaac Newton, *Principia*, Vol. I: *The Motion of Bodies*, 13.

admit also of the existence of an interacting force—whatever it may be—that keeps the dispersed particles or mass particles together.

All entities exert a reciprocal force upon one another, and in so doing, keep the material universe together! This assertion is rather speculative, yet Newton couches his overall view of connection and interaction in a precise mathematical formula: "the mutual actions of two bodies upon each other are always equal and directed to contrary parts." He explains:

> Whatever draws or presses another is as much drawn or pressed by that other. If you press a stone with your finger, the finger will also be pressed by the stone. [...] If a body impinge upon another, and by its force change the motion of the other, that body also, (because of the equality of the mutual pressure) will undergo an equal change in its own motion, towards the contrary part.[23]

From the background against which this law must be read—the inter-action of bodies or particles moving in a *void*—it is clear that what Newton sought to convey is the phenomenon of "mutual attraction in spite of the distance." This attraction can take on various modalities, though the basic structure is the same. Newton tells us, e.g., that the "gravitation between the Earth and its parts is mutual;" the force which the Earth as a whole impinges on its parts is equal to the contrary (composite) force its parts forcefully impinge on it. If there would be a non-equivalence in this matter, the whole Earth, which is floating in unresisting ether, would be put off balance, and "be carried off *in infinitum*." [24] In fact, the "mutual attraction" of the Earth and its parts is not that spectacular, for although there is a void between the mass particles in the womb of the Earth, the very bulk of the Earth and the very bulk of all its particles still occupy the same spatial area.

Things will be significantly different in the case of two celestial bodies A and B put at a considerable distance from each other. Here, too, a mutual attraction will take place. All the mass particles of celestial body A are going to act forcefully upon those of celestial body B, and reciprocally, all the mass particles of the latter are going to act forcefully upon the former. This time, however, the mutual action of the two bodies upon each other is more dramatic. The forces now literally direct themselves "to *contrary parts*" (to parts located elsewhere in empty space) with the result that they

---

23   Ibid., 13-14.

24   Ibid., 26.

tend to fly "into each other's arms" in spite of their spatial distance. This is Newton's initial intuition of the phenomenon of gravitational attraction.

When commenting on the difference between Newton and Descartes, Alexandre Koyré refers to the witty remark of Voltaire, who wrote that

> a Frenchman who arrives in London finds himself in a completely changed world. He left the world *full*; he finds it *empty*. In Paris the universe is composed of vortices of subtle matter; in London there is nothing of that kind. In Paris everything is explained by pressure which nobody understands; in London by attraction which nobody understands either.

Koyré goes on:

> Voltaire is perfectly right. The Newtonian world is chiefly composed of a void. It is an infinite void, and only a very small part of it—an infinitesimal part—is filled up, or occupied, by matter, by bodies which, indifferent, and unattached, move freely and perfectly unhampered in—and through—that boundless and bottomless abyss. And yet it is a world and not a chaotic congeries of isolated and mutually alien particles. This, because all of these are bound together by a very simple mathematical law of connection and integration—the law of mutual attraction—according to which *every one of them is related to and united with every other*. Thus each one takes its part and plays its role in the building of the system of the world.[25]

The Universal Law of Gravitation

Its Formula

The general principle of this law stipulates

> that "there is a power of gravity pertaining to all bodies, proportional to the several quantities of matter they contain." The universe is composed of particles of matter all of which attract each other with a force proportional to the products of their masses and inversely proportional to the square of the distance between them.[26]

Applied to the Earth and the Sun, this means that the force that acts attractively between them has a strength that is proportional to the product of the Earth's and the Sun's masses and inversely proportional to the square of the distance between their centers.

Encouraged by the young astronomer Edmund Halley, who in 1684 pressed him to undertake this research, Newton began to apply his laws of

---

25   Alexandre Koyré, *Newtonian Studies*, 14.
26   Richard Westfall, *The Construction of Modern Science*, 155.

motion to the revolution of the planets. He needed this challenge because, at that moment, two years before the publication of the monumental *Principia Mathematica*, he was recovering from a state of depression. Newton thus resumed a research topic that had kept him busy in earlier days but which, for several reasons, he had dropped. Looking back now at these earlier reflections—Newton's written notes are all conserved—one notices that, at that time, his analysis of the Moon's revolution about the Earth had not brought him considerably further than the insights of his contemporaries. Although his study of magnetism occasioned him to experiment with the "inverse square law" to calculate the declining propagation of forces in the void, he did not yet apply this phenomenon to the mutual *attraction* between two celestial bodies; he still thought that the Moon, in its fall movement, was constantly rushing towards the Earth but failed to hit the Earth because of the *centrifugal* force that swept it away. This prominence given to the centrifugal force was something he carelessly had been sharing with the Cartesian school and especially with the Dutchman Huygens.

At the occasion of a biting discussion with Robert Hooke, who was then secretary of the Royal Society, Newton came to espouse Hooke's suggestion that the orbital deflection of the Moon—and of every planet— was due to *central attraction*. This change in perspective was decisive.

> Hooke's suggestion of a central attraction came exactly at the time when Newton's speculations had led him to assert the existence of forces between particles. He was in a position to accept the idea of attraction as he had not been before. The idea of attraction, in turn, offered physical content to the mathematical abstraction of force toward which his earlier work in mechanics (from 1660 till 1666) had moved. In a word, all of the factors were now present to produce the concept of universal gravitation.[27]

Starting with Book I of the *Principia Mathematica*, Newton coined the new term *centripetal force*, a force that seeks the centre, in deliberate contrast to Huygens' term *centrifugal force*. He defines it as "the force by which bodies are drawn or impelled, or any way tend, towards a point as to a centre," and he adds the following explanation:

> Of this sort is gravity, by which bodies tend to the centre of the Earth; magnetism, by which iron tends to the loadstone; and that force, *whatever it is*, by which the planets are continually drawn aside from the rectilinear motions, which otherwise

---

27   Ibid., 150.

they would pursue, and made to revolve in curvilinear orbits. A stone, whirled about in a sling, endeavors to recede from the hand that turns it; and by that endeavor, distends the sling, and that with so much the greater force, as it is revolved with the greater velocity, and as soon as it is let go, flies away. That force which opposes itself to this endeavor, and by which the sling continually draws back the stone towards the hand, and retains it in its orbit, because it is directed to the hand as the centre of the orbit, I call the centripetal force. And the same thing is to be understood of all bodies, revolved in any orbits. They all endeavor to recede from the centers of their orbits; and were it not for the opposition of a contrary force which retains them to, and detains them in their orbits, which I therefore call centripetal, they would fly off in right lines, with a uniform motion.[28]

Later on, he becomes even more specific: "The force which retains the celestial bodies in their orbits has been hitherto called centripetal force; but it being now made plain that it can be no other than a gravitating force, we shall hereafter call it gravity."[29]

The formula of the law of gravitation is simple and concise: $F = \frac{Mm}{r^2}$. That one has to multiply the masses of the Earth (m) and the Sun (M)—and not just some parts of them—is evident, for that product expresses the *mutual* force (F) of the two massive bodies when they act upon each other in contrary directions and invest the whole bulk of their respective quantities of matter in that mutual gravitational pull. This rule of thumb, however, is manifestly incomplete, and has to be supplemented with the 'inverse square law' of the distance. Since the mutual gravitational attraction operates through free space, its range or propagation of force declines in proportion to the widening spatial interval: the force varies inversely to the square of the distance (r). The "inverse square law" comes to the fore, first, in the planets' varying year of revolution about the Sun, and second, in the varying speeds they adopt on their respective elliptical orbits.

Because Neptune, for example, revolves farther from the Sun, it is less exposed to the latter's attractive power than, say, Mercury, which is nearer and is also a smaller planet. Neptune's specific year of revolution is consequently longer than that of Mercury. Taken in itself, Neptune's bigger mass could make it gravitate much swifter towards the Sun than Mercury, but this inherently stronger force is overridden by its considerably farther distance, which allows it to "escape" the mighty gravitational pull, whose force declines with the increasing square of the distance ("inverse square

---

28   Isaac Newton, *Principia*, Vol. I: *The Motion of Bodies*, 2–3 (italics mine).

29   Isaac Newton, *Principia*, Vol. II: *The System of the World*, 410.

law"). The escape is, of course, never complete since the force of gravitation continues to retain the planet in its orbit, and compels it to stay there and neither to fly off '*in infinitum'* nor to adopt the rhythm of a uniform right line motion to which it would return without the influence of the force.

Besides this aspect, there is also each planet's oscillating speed on its specific elliptical orbit, a speed that also obeys the "inverse square law" of the distance. Indeed, in its *aphelion* (farthest distance from the Sun), the planet moves slower than in its *perihelion* (closest distance to the Sun) in the vicinity of which it accelerates its speed. Whereas Kepler was already able to relate this rhythm of acceleration and deceleration to the changing length of the vector radius, Newton succeeded in reformulating this insight from the perspective of gravitation. Mercury, for instance, is drawn by the gravitational pull most intensely at its perihelion. There, the planet is attracted more forcibly towards the Sun, and in a sense, falls swiftly towards it. Once this perilous cape is left behind, however, the planet sets out to swing back, in an elongated curve, towards its slow motion zone near the aphelion. The transition from *perihelion* to *aphelion* accounts for the elliptical form of the planet's orbit.

Rebuttal of Descartes' Vortex Theory

Newton came to his central notion of gravity through a meticulous analysis of physical phenomena but also through a reflection on the nature of matter and space, a topic about which he significantly diverged from Descartes. In order to refute Descartes, Newton set out to *mathematically* demonstrate that his picture of the world was wrong. In book II, Sec. IX of the *Principia mathematica* he probed into the physical consequences of Descartes' vortex theory.

In reconstructing the vortex mechanism, Newton carefully examined the wave-like propagation of kinetic impulses in fluids. A stone, for example, that is thrown into a lake will produce spreading rings that gradually move away from the centre. So too, a rotating cylinder sunk into deep waters will force the surrounding masses of water to whirl around in spreading concentric circles. Of these gyrating concentric masses of water one can measure the periodic times, i.e., the times needed for the separate parts of the fluid to whirl about the axis of the cylinder. These periodic times turn out to be in the same degree as the distances from that axis.

Newton imagined a cylinder that is infinitely long:

> If a solid cylinder infinitely long, in an uniform and infinite fluid, revolves with an
> uniform motion about an axis given in position, and the fluid be forced round by

only this impulse of the cylinder, and every part of the fluid continues uniformly in its motion: I say, that the periodic times of the parts of the fluid are as their distances from the axis of the cylinder.[30]

Thus, if the distance from the axis is 5 miles, the periodic times will also be 5 time units; if the distance is 6 miles, the periodic times will be 6 time units.

To typify the vortex mechanism, one ought to substitute a gyrating solid sphere for the rotating solid cylinder and keep all the other parameters unchanged. In this case too, the solid sphere forces the fluid to whirl and spread the rotating parts in expanding circles. There is, however, a remarkable difference: the times needed for the parts of the fluid to whirl round the centre of the solid sphere are in the same degree as the squares of the specific distances from that centre. The squaring flows from the fact that the rotating globe, unlike the infinitely long solid cylinder, freely floats in the fluid, so that whirling parts of the fluid also pass above and underneath the rotating globe. In Newton's words:

> If a solid sphere, in an uniform and infinite fluid, revolves about an axis given in position with an uniform motion, and the fluid be forced round by only this impulse of the sphere; and every part of the fluid continues uniformly in its motion: I say, that the periodic times of the parts of the fluid are as the *squares* of their distances from the centre of the sphere.

He further explains:

> For the central rotating solid sphere] will communicate a whirling motion to the fluid like that of a vortex, and that motion will by degrees be propagated onwards *in infinitum*; and this motion will be increased continually in every part of the fluid, till the periodical times of the several parts become as the squares of the distances from the centre of the globe.[31]

If the distance from the axis is 5 miles, then the periodic times will be 25 time units; if the distance is 6 miles, the periodic times will be 36 time units. This means that the gyrating motion is considerably slowing down the farther it recedes from the centre.

---

30   Isaac Newton, *Principia*, Vol. I: *The Motion of Bodies*, 385.

31   Ibid., 387, 390.

Having set this stage, Newton proves that the vortex mechanism cannot be applied to the solar system. If the solar system were to work that way, it would gradually disintegrate, for on close inspection, the whirling motion is purely centrifugal. A centrifugal motion that is by degrees propagated onwards *in infinitum* and is continually squaring its periodic times with the growing distance must of necessity take on a slower rhythm of circulation as it stretches towards the periphery. The motions of the whirling parts will languish by degrees, and the vortex at last will stand still. Even if one were to assume that the universe is made up of innumerable vortices that keep one another in balance by their reciprocal pressures (a position held by Descartes), it is clear that a resulting equilibrium is very unlikely. Vortices that are intrinsically threatened by exhaustion—such being their common predicament—cannot possibly stabilize one another. Newton's point is clear: if there is no gravitational force that keeps the planets on their orbits, our solar system and all other solar systems are doomed to disintegration.

More importantly, Descartes' system clashes with Kepler's third law. This law states that

> the *squares* of the periods of the planets (the times for them to complete one orbit) are proportional to the *cubes* of their average distance from the Sun; the more distant the planet, the more slowly it moves, but according to a precise mathematical law: $P^2 = a^3$; where P presents the period of revolution of the planet around the Sun, measured in years, and a the distance of the planet from the Sun measured in "astronomical units" (an astronomical unit is the distance of the Earth from the Sun). Jupiter, for example, is five astronomical units from the Sun, and 5 x 5 x 5 = 125. What number times itself equals 125? Why, 11 is close enough. And 11 years is the period for Jupiter to go once around the Sun.[32]

Now, this ratio of cubes and squares does not fit into Descartes' vortex system with its swelling squares of periodic times. Jupiter accordingly would not have a periodic time of 11 years but of 25 years (5 x 5). A planet at a distance of 6 astronomical units from the Sun would not have a periodic time of 14.5 years (6 x 6 x 6 = 216, which is roughly equivalent to **14.5** x 14.5 = 210) but of 36 years; a planet at a distance of 7 astronomical units would not have a periodic time of 18.5 years (7 x 7 x 7 = 343, which is roughly equivalent to **18.5** x 18.5 = 342) but of 49 years.

---

32   Carl Sagan, *Cosmos* (New York: Random House, 1980), 63.

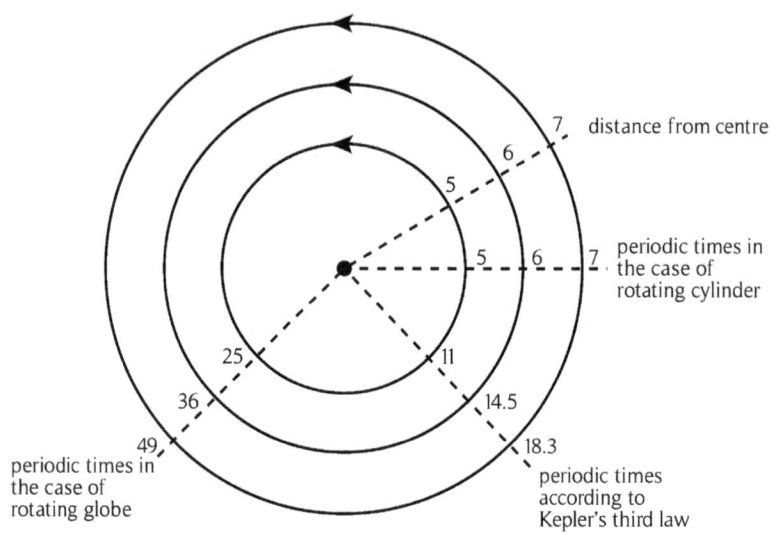

Figure 13. Varying Periodic Times

Kepler's ratio of cubes and squares perfectly agrees with Newton's universe whose coherence hinges on the centripetal gravitational attraction *spreading through the void.*

> It must be granted, so being mathematically demonstrated, that, if celestial bodies revolve with an equable motion in concentric circles [ellipses respectively], and the squares of the periodic times are as the cubes of the distances from the common centre, the centripetal forces will be inversely as the squares of the distances.[33]

True, in the above description one has also to do with a propagation of forces that decline with the increasing square of the distance; yet the effect of this squaring is different from that obtained from the spread of motion *in the medium of a fluid.* In Newton's approach, there is from the outset a mighty pull that draws bodies towards the centre and whose force of attraction—which pierces the void—declines with the square of the distance. This decline, however, cannot be likened to Descartes' vortex in which the contiguous whirling parts languish by degrees the more they are spreading to the periphery (here, quantities of motion are being subtracted from quantities of motion). In Newton's approach, the first mighty pull,

---

33   Roger Cotes, *Preface,* in Isaac Newton, *Principia,* Vol. I: *The Motion of Bodies,* xxii.

which is short-ranged, adds itself to the whole cascade of lesser pulls that follow, and the sum total of it does not result in the exhaustion of the system but in a powerful attraction with a long-range effect that varies inversely to the square of the distance and whose effect on the periodic times of the planets agrees with Kepler's third law. Kepler's third law, formulated 75 years before the publication of the *Principia Mathematica*, settles the dispute as to whether the planets are kept in their orbits on the basis of a centrifugal or a centripetal force, and gives the victor's palm to Newton. Kepler's ratio of the 3/2nd power only applies to a universe in which the centripetal force of gravity is constantly at work. "Let philosophers then see," Newton ironically remarks, "how that phenomenon of the 3/2nd power can be accounted for by vortices."[34]

The result of Newton's discussion with Descartes is rather stupendous. In rebutting his opponent, Newton made it clear that, for him, the force of gravity, which retains the planets in their orbits, is a totally *immaterial* force that acts at a distance through the *void*. Only as an immaterial entity can the force of gravity behave as it does: propagating its gravitational pull inversely to the square of the distance from the centre.

Newton's Doctrine of God

Newton abhorred the Cartesian equivalence of matter and extension because he feared that sooner or later it would give rise to a purely materialistic conception of the world. For him, "the Cartesian idea that the well-ordered system of the world could be the result of mere mechanical causes is absurd. The most beautiful system of Sun, planets, and comets, could only proceed from the counsel and dominion of an intelligent and powerful Being."[35] This intelligent and powerful Being must exert its dominion *always* and *everywhere*. For Newton this insight is so overwhelming that he associates God with absolute space and absolute time. Absolute space and absolute time form the mathematical framework that gives consistency to all the motions in the universe; they are also the abode of the infinite and omnipresent God, the sanctuary from which God forcibly acts upon the world.

---

34   Isaac Newton, *Principia*, Vol. I: *The Motion of Bodies*, ed. Florian Cajori, 394.

35   Alexandre Koyré, *Newtonian Studies*, 111.

Absolute Space and Time

Newton reckoned with the existence of several solar systems, and placed our solar system in the middle of them. In line with ancient cosmologies, he sought, first of all, to give a solid foundation to the solar system by stating that the "common centre of gravity of the Earth, the Sun, and all the planets, is immovable."[36] The Sun, which is close to this common centre of gravity, basically perseveres in its state of rest. If it is slightly rocking, this is the effect of its gravitational interaction with the rest of the point masses in the solar system. On the whole, however, the Sun "never recedes far from the common centre of gravitation of all the planets."[37]

In line with his conception of the void, Newton, second, posited that the biggest concentration of stars and clusters of stars is to be found near the solar system, whereas farther away, these clusters seem to be "sown" more sparsely. The more one approaches the periphery of the universe, the fewer stars one is going to encounter, and the greater also the void grows. In Newton's universe, "the stellar universe ought to be a finite island in the infinite ocean of space."[38] Compared with this ocean, the clusters of stars are but minuscule "point masses," for the most part located in one small niche of universe, and surrounded by the immensity of space that tremendously surpasses them. The same is true for the solar system. Here, too, the areas of empty space by far outnumber the areas filled with matter. Wherever one were to place oneself, one would always observe that the universe is largely empty.

The existence of the void is crucial to Newton. Only when the universe is conceived as an immense "void" within which material bodies can interact with one another at a distance is there a place left for God. No wonder that he spent so much energy in demonstrating the astronomical absurdities that flow from Descartes' vortex theory:

> Newton's anti-Cartesianism is not purely scientific; it is also religious; Cartesianism is materialism that banishes from natural philosophy all teleological questions, that reduces everything to blind necessity, which obviously cannot explain the variety and the purposeful structure of the universe.[39]

---

36   Isaac Newton, *Principia*, Vol. II: *The System of the World*, ed. Florian Cajori, 419.
37   Ibid.
38   Albert Einstein, *Relativity. The Special and the General Relativity: A Popular Exposition*, trans. Robert Lawson (New York: Bonanza Books, 1961), 106.
39   Alexandre Koyré, *Newtonian Studies*, 109.

And where there is no purposeful structure, there is no planning or acting God.

Having stressed the importance of empty space, Newton sets out to mathematically define the notions of absolute space and absolute time. He, first of all, makes it clear that space, is an independent receptacle within which the particles of matter—the aggregates of point masses, such as stars and planets—perform their ever-regular motions and accelerations. Second, he spells out that only on condition that the spatial container itself is never subject to change can the various changes in the universe be organized purposefully. Only the objective coordinates of absolute, unchangeable space and absolute, unchangeable time can "warrant the validity of the mathematical constructions of physics."[40] Space and time must be absolute values, if the laws of nature are to work without any exception.

For Newton, absolute space and absolute time differ from our ordinary, "relative" experiences of space and time. He writes:

> Absolute space, in its own nature without relation to anything external, remains always similar and immovable. Relative space is some movable dimension or measure of the absolute spaces; which our senses determine by its position to bodies; and which is commonly taken for immovable space; such as the dimension of a subterranean, an aerial, or celestial space, determined by its position in respect of the Earth.[41]

In other words, what we consider an allegedly immovable space, such as the spatial zone of the fixed stars, is in fact but a remote likeness of what absolute, immovable space, in and of itself, is. Why? The reason is quite simple: we always perceive this "apparent" immovable space from the standpoint of our Earth, and thus, relative to the Earth's rotation. Our own changing perception makes us always miss the very core of absolute space: its *lack of relation whatsoever to anything external*.

The same is true for absolute time.

> Absolute, true, and mathematical time, of itself, and from its own nature, flows equably without relation to anything external, and by another term is called duration; relative, apparent, and common time, is some sensible and external [...]

---

40   Jacques Merleau-Ponty and Bruno Morando, *Les trois étappes de la cosmologie* (Paris: Laffont, 1971), 94 (translation mine).

41   Isaac Newton, *Principia,* Vol. I: *The Motion of Bodies,* ed. Florian Cajori, 6.

measure of duration by the means of motion, which is commonly used instead of true time; such as an hour, a day, a month, a year.[42]

Absolute time endures from eternity to eternity (or as Newton writes, from infinity to infinity), whereas we, from our changing place in the universe, are able to grasp absolute time only by channeling it into our relative estimation of sequences of time. We do such channeling when inferring the demarcation of months and years from the changing astronomical positions of celestial bodies. Yet, since these changing positions are always relative to one another; they can never disclose the absolute, true time that flows equably without relation to anything external.

For Newton, all the motions in the universe are related to and contained within the absolute coordinates of space and time, whereas these coordinates themselves have no relation whatsoever to these motions. These absolute coordinates are not in the least influenced by the various motions in the universe, whereas only within these coordinates can temporal motions acquire their intelligible character. This tenet calls to mind the classical definition of God that can be found in Aquinas: "The creatures have a proper relationship to God, whereas in God there is no proper relationship to the creatures."[43] It was this religious truth that Newton apparently sought to intimate with his considerations on the absolute coordinates of space and time. He was convinced that there must be an affinity between God's eternity, infinity, and immutability on the one hand, and the eternity, infinity, and immutability of absolute space and time on the other.

The Disclosure of Absolute Space and Time in the Regularity of Absolute Motions

In a further move, Newton examines which kind of motion may indirectly give us an idea of the basic notions of absolute space and absolute time. In this light, he draws a distinction between absolute motion and relative motion and defines them as follows: "Absolute motion is the translation of a body from one absolute place into another [absolute place]; and relative motion is the translation from one relative place into another [relative place]." Both translations may happen at the same time, but even then, the translation from absolute place to another absolute place turns out to be the most significant. He gives the example of a sailor who walks with a velocity of one unit eastward on the deck of a ship that "itself,

---

42   Ibid.
43   Thomas Aquinas, *Summa theologiae*, I, q. 13, art.8.

with a fresh gale, and full sails, is carried towards the west with a velocity expressed by 10 units," whereas the Earth, in whose waters the ship sails, at that moment rotates eastward about its axis with a velocity of 10,010 units. Thus, the sailor is moving both according to a relative motion and an absolute motion. "The sailor will be moved truly in immovable space towards the east, with a velocity of 10,001 parts [units] and relatively on the earth towards the west, with a velocity of 9 of those parts [units]."[44]

The example illustrates what Newton had in mind. The sailor moves at a velocity of 10,001 units eastward: the velocity of walking on the deck (+1) added to the velocity of the rotating Earth, which also goes east (+10,010), minus the velocity of the ship sailing west (-10). Now, this translation eastward makes him move truly in "immovable space" in spite of the fact that he is carried by the ship into the opposite direction. One would have expected Newton to name this translation "*absolute motion*" (measured by the earth's unchanging rhythm of motion). Instead, he declared that the sailor moves with this velocity "truly *in immovable space* towards the east." This is, of course, no contradiction but only a hasty manner of jumping to a conclusion. Indeed, according to the given definition, the Earth's steady rhythm of motion can obviously be called an "absolute motion," i.e., a motion through which a celestial body is carried from one absolute *place* into another absolute *place*, but in order to grasp how absolute *places* relate to absolute *space*, some extra information is needed which he gives a few pages further. There it is said: "No other places are immovable than those that, from infinity to infinity, do all retain the same given position one to another; and upon this account must ever remain unmoved; and do thereby constitute immovable space."[45]

Thus, absolute or immovable *space* is "made up" of absolute or immovable *places* whose given positions endure from eternity to eternity. To visualize what this means, one has to abstract from sense impressions and rely on one's mathematical imagination. One has to "imagine" that the Earth on which the ship sails (and on which, in turn, the sailor walks) is rotating about its axis and about the Sun in a rhythm that is never going to change—thus allowing itself to be carried from "immovable place" to "immovable place" along an orbit whose fixed points of reference are never going to change. If one would ask why these fixed points of reference are never going to change, the answer is obviously: because they are anchored

44   Isaac Newton, *Principia*, Vol. I: *The Motion of Bodies*, 7.
45   Ibid., 9.

in the overarching immutability of absolute space and time, in which the planet's never changing rhythm of motion partakes.

The planet's steady rhythm of motion *is*, of course, not the absolute immovable space but only takes place *in* that absolute space that endures from eternity to eternity. Newton seems to intimate that a planet's ever identical rhythm of motion is, so to speak, the "sacrament," the visible sign and symbol of the self-presentation of *the* absolute space that, along with absolute, mathematical time, constitutes the ultimate *measure* of ordered motion in the universe. This ultimate measure of which we have no adequate intuition discloses itself in glorious manifestation whenever we experience the clashing contrast between absolute and relative motions. At that moment, it dawns upon us how much more dignity there is in the constant rhythm of revolution of each of the planets than in the ordinary straight-line movements. This contrast experience allows us to perceive the cosmic revelation of the absolute. The absolute co-ordinates and God, in the deity's infinity and tremendous majesty, manifest themselves in the ordered spectacle of the celestial rotations—a spectacle that fills us with awe.

Compared with the disclosing power of the regularly rotating planets, the uniform, straight-line movements, whether accelerated or decelerated, are just ordinary motions bereft of excellence. The ship sailing west with a velocity of 10 units on the earth's globe, which is turning east, represents, in comparison, an unimpressive movement. Its cruise has no impact whatsoever on the planet's ever-identical rotation, which is not in the least hindered by the ship making headway in the opposite direction. Whether or not the ship is there or speeds up its velocity, the Earth keeps moving in its unaltered rhythm of rotation, from year to year, from century to century, from millennium to millennium. By contrast, the sailor who is taking some steps in the direction of the Earth's rotation seems to join forces with the globe. He sees his steps subtracted from the velocity with which the ship is carrying him west. His stroll on the deck makes him "participate" in the Earth's cosmic rotation.

Did Newton slide back into the ancient admiration of the revolving celestial bodies? The answer is "yes" and "no." Yes, in so far as he attached great importance to the regular, quasi-circular revolution of the planets. This brought him in the company of Plato and Aristotle, who used to ascribe a particular excellence to the circular motion, in contrast to the ordinary straight-line motions in the sublunary sphere. For them the rotating circle was the visible manifestation of the deity. The celestial bodies constituted an object of worship because of their most perfect circular orbits. To contemplate these gracefully revolving forms in the sky was to experience eternity in the midst of temporality. It is evident that Newton revisited the classical Greek tradition and its Christianized variants; yet this retrieval was by no means a mere repetition. Newton transposed the cosmic religiosity into the parameters of the heliocentric system and the celestial physics on which it rests. That makes his contribution highly distinct from his Greek predecessors. His probing into the laws of physics could only make him smile at the curious constructions of his ancient predecessors. The heliocentric universe needs neither crystalline hoops to make the stars and planets revolve, nor a host of cosmic intelligences under God to harmonize the revolutions. The only immaterial force that, next to God, is formative of the planets' steady orbiting from absolute place to absolute place is the force of gravity.

The Force of Gravity: "Secondary Cause" in Creation

Although Newton did not use this term, it is clear that, for him, the force of gravity was a 'secondary cause' the Creator needed to give a solid structure to the universe. With the help of this force the Creator God sees to it that the celestial bodies are retained on their trajectories as they move from one absolute place into another within the co-ordinates of absolute space and time. Moreover, this force is a force with a universal impact: No entity in the universe can escape its vigorous pull. Could the Creator have chosen any better instrument for enjoining obedience upon creatures, for assigning to them their proper place?

Newton was apparently so impressed by the effects of the force of gravity that he began to regard it as a wondrous power whose inner working he was not able to understand. He wrote: "Hitherto we have explained the phenomena of the heavens and of our sea by the power of gravity, but have not yet assigned the cause of this power."[46] In other words, we are able to couch the effects of this force in a precise mathematical formula, but we are not able to know its true nature which remains shrouded in mystery. Yet, the undeniable fact of its potent existence can be demonstrated through experiments: "For instance, if two globes, kept at a given distance one from the other by means of a cord that connects them, were revolved about their common centre of gravity, we might, from the tension of the cord, discover the endeavor of the cord to recede from the axis of their motion, and from

---

46   Isaac Newton, *Principia,* Vol. II: *The System of the World,* 546.

thence we might compute the quantity of their circular motions,"[47] a quantity that reveals the effect of the gravitational pull.

From experiments like this some scientists deduced that gravity was an inherent quality of matter. Yet, Newton firmly rejected this view. In a letter to Richard Bentley, dated January 17, 1693, he wrote: "You sometimes speak of gravity as essential and inherent to bodies. Pray do not ascribe this notion to me, for the cause of gravity is what I do not pretend to know and therefore would take more time to consider it."[48] For Newton, gravity must be a totally immaterial force; if not, it could not be used by the immaterial Creator God to impose his patterns of order on the material world. The focus on this immaterial character comes to the fore in a subsequent letter to Mr. Bentley, dated February 25: "It is inconceivable," Newton wrote, "that inanimate brute matter should, without the mediation of something else which is not material , operate upon and affect other matter, without mutual contact."[49] For him, it is clear that the mysterious force of gravity is an instrument in God's hands and not just an inherent property of matter itself. The mysterious force of gravitation reveals the power and dominion of the omnipotent God.

The Might of God Omnipotent Acting in the Medium of Absolute Space and Time

In classic Christian theology, boundless space and time were regarded as attributes of God; as such, they denoted a space and time that were of another quality than the same notions in the finite world. The notions immensity, infinity, and eternity were used to highlight the divine sovereignty that endures forever and ever. Compared with God, the eternal one, all the metaphors describing the vastness and steadiness of the heavens were bound to disappear into naught. Yet, in Newton's days, the notions of infinity and eternity began to be used to designate the mathematical coordinates of absolute space and time. Divine attributes were transposed to the spatio-temporal coordinates of the universe.

Facing up to this "transfer" of attributes, Henry More (1614–1684), a renowned Christian neoplatonist in Cambridge, set out to harmonize the mathematical concepts of absolute space and time with God's infinity. He made it clear that the material components of the universe are finite and

---

47   Isaac Newton, *Principia*, Vol. I: *The Motion of Bodies*, 12.

48   H. Thayer, *Newton's Philosophy of Nature: Selections from His Writings* (London: Hafner Press, 1974), 53.

49   Ibid., 54.

created, whereas the absolute space in which these material elements are contained is uncreated and infinite; it shares in the Creator's attributes of immobility and eternity: "Absolute space is eternal and therefore uncreated. But the things that are in space by no means participate in these properties. Quite the contrary; they are temporal and mutable and are created by God in the eternal space and at a certain moment of the eternal time."[50] In an anti-Cartesian mood, More not only separated space from matter, but raised the infinite void of space "to the dignity of an attribute of God, and of an organ in which and through which God creates and maintains His World."[51]

This state of affairs raises the question as to which of the following infinity and necessary existence pertain: to the sole Creator God or to the Creator God plus his two, co-eternal "cosmic hands," absolute space and time? Are infinity and eternity God's sole prerogative or are they also characteristics of the universe's spatio-temporal co-ordinates? More did not draw back "before the conclusions of his premises with which he announced to the world the spatiality of God and the divinity of space."[52]

This fluctuation in the attributions of infinity and eternity can also be found in Newton. In passages in which he adopted the style of the Hebrew Bible, God's infinity, omnipresence, and omnipotence are strictly seen as the deity's sole characteristics, whereas elsewhere, in the wake of More's Neoplatonism, he elevates absolute space and time to the *sensorium* God uses to make his divine omnipresence felt in creation.

The Hebrew influence comes to the fore in Newton's confession of faith in the General Scholium:

> This most beautiful system of the Sun, planets, and comets, could only proceed from the counsel and dominion of an intelligent and powerful Being [...] This Being governs all things, not as the soul of the world, but as Lord over all; and on account of his dominion he is wont to be called *Lord God*, *pantokratoor*, or *Universal Ruler* [...] The supreme God is a Being eternal, infinite, absolutely perfect; but a being, however perfect, without dominion, cannot be said to be Lord God [...] And from his true dominion it follows that the true God is a living, intelligent, and powerful Being [...] He is eternal and infinite, omnipotent and omniscient; that is, his duration reaches from eternity to eternity; his presence from infinity to infinity; he governs all things, and knows all things that are or can be done. He is

---

50   Alexandre Koyré, *From the Closed World to the Infinite Universe*, 150.

51   Ibid., 152–53.

52   Ibid., 152.

not eternity and infinity; but eternal and infinite; he is not duration and space, but he endures and is present. He endures forever, and is everywhere present; and by existing always and everywhere, he constitutes duration and space.[53]

The Creator is, thus, the first sovereign cause of duration and space, i.e., of the immaterial receptacle in which he calls forth the material elements and their motions. Absolute space and time are themselves creatures, constituted by God, and not emanations flowing from his eternal Being.

Yet, there is another famous passage in the *Opticks,* in which Newton, in the style of the theophany in the book Job, asks the question as to where all the purposefulness in the world does come from. It is in this context that one finds an allusion to Henry More's idea of infinite space being the omnipotent God's "organ" through which He perceives and steers the entities in the world:

> What is there in places almost empty of matter, and whence is it that the Sun and the planets gravitate towards one another, without dense matter between them. Whence is it that nature does nothing in vain, and whence arises all that order and beauty that we see in the world? [...] How came the bodies of animals to be contrived with so much art, and for what ends were their several parts? Was the eye contrived without skill in optics and the ear without knowledge of sounds? [...]And these things being rightly dispatched, does it not appear from phenomena that there is a Being, incorporeal, living, intelligent, omnipresent, who in infinite space, as it were in his sensorium, sees the things themselves intimately, and thoroughly perceives them, and comprehends them wholly by their immediate presence to himself?[54]

The German philosopher Gottfried Leibniz, who apparently could not omit picking a quarrel with Newton, comments on the above text as follows: "Sir Isaac Newton says, that Space is the Organ, which God makes use of to perceive things by. But if God stands in need of any Organ to perceive things by, it will follow, that they do not depend altogether upon him, nor were produced by him."[55] This interpretation is, of course, biased, and calls to mind the objection Leibniz earlier raised to Henry More's view: that his notion of absolute space is a duplicate of God, another divinity standing over and against divinity. To do justice to Newton, however, one

---

53   Isaac Newton, *Principia*, Vol. II: *The System of the World,* 544–45.

54   Isaac Newton, *Opticks,* ed. Bernard Cohen (New York, 1952), 369.

55   Alexandre Koyré, *From the Closed World to the Infinite Universe,* 235.

has to examine carefully the message he sought to convey with the above quoted passages. Newton, first of all, affirms that God, by existing always and being omnipresent, constitutes absolute space and time, and, second, that having constituted this encompassing framework, God makes use of it as his *sensorium* by which to perceive things and to steer them towards their purposeful actions. Leibniz deliberately put "organ" to underline Newton's indebtedness to More. That is the reason why Dr. Clarke, a friend of Newton, specified on his behalf that *sensorium* is to be understood as "place of sensation": "The word sensory does not properly signify the organ but the place of sensation. The eye, the ear, etc. are organs, but not sensori-al."[56] This implies that the Newtonian God is not in an animistic sense, the soul of the world, nor that God,—as Leibniz imputed to Newton,—would stand in need of an organ of perception not created by the deity. This clar-ification, however, did not resolve the theological discussion with which Leibniz sought to entrap his opponent.

The correspondence between Leibniz and Clarke in the years 1715 and 1716 was published in London in 1717. In it, Leibniz defended his deistic conception: God has from the outset planned all the details of the world so perfectly that He no longer needs to intervene in the world's workings. Clarke, on the other hand, tried to justify Newton's view that God as the sovereign ruler acts upon the world through the spatial framework He has intentionally called forth. Clarke highlights that

> God is not in space or in time, but his existence is the cause of space and time. And
> when we say, in conformity with the language of the vulgar, that God exists in all
> the spaces and in all the times, these words mean only that he is omnipresent and
> eternal, that is, that boundless space and time are *necessary consequences* of his
> existence, and not that space and time are beings distinct from him, and in which
> he exists.[57]

Clarke made it clear that absolute space and time are not separate entities disconnected from God (this was precisely what Leibniz hinted at in his criticism), but that this connection must be understood in the manner of Plotinus: boundless space and time are the *necessary conse-quences* of God's existence. In this light, Leibniz's trap became successful, for in the neoplatonic tradition, the fine-tuned mathematical connections of things (the world of ideas) figured as the first emanation from the One,

---

56   Ibid., 242.

57   Ibid., 271 (italics mine).

an emanation that necessarily flows from the deity's Being as soon as the intellect (the cosmic Nous) seeks to understand the simple being of the One. The 17th-century Oxford neoplatonists apparently replaced the world of ideas with the absolute coordinates of space and time, while for the rest upholding the emanation model. Yet, in doing so they espoused the necessity with which the emanations flow from the One. Indeed, in Neoplatonism, the emanations are logical and not free, and in that sense, are of necessity overflows from God's eternal Being. On the other hand, one ought not to forget that in Plotinus' system of emanations the sovereignty of the supreme Being is never at stake, since the One always remains elevated above the things that through emanation share in his divine properties. This is what Clark intends to say when stating that "God is *not* in space or in time, but his existence is the cause of space and time." The superabundance of the ultimate source radically transcends the coordinates of mathematical space and time it calls into existence by way of necessary emanation.

CHAPTER 8

Einstein

Albert Einstein (1879–1955) revolutionized the human understanding of the universe and of our place in it with his two daring assumptions, viz., the special (or restricted) theory of relativity (1905) and the general theory of relativity (1915).

Special Theory of Relativity

In order to grasp Einstein's revolutionary approach one ought to realize the basic problem that confronted the scientific community at the end of the 19th century: the incompatibility of the newly discovered laws of the electromagnetic field and classical Newtonian mechanics. In the mid 19th century, considerable research had been done in the domain of magnetic and electric fields by Michael Faraday (1791–1867) and James Maxwell (1831–1879). The results of their research no longer fit into the general framework of Newtonian physics. Newton had focused his attention exclusively on the motions of mass particles within the unchangeable coordinates of space and time. Yet, the electromagnetic fields behaved differently from mass particles in that they propagate themselves through undulations with a speed that equals the speed of light (300,000 km a second); moreover they carry a determinate amount of energy with them. The precise connection Maxwell established in 1865 between electromagnetic fields and light waves was one of the great accomplishments of 19th century mathematical physics, although it was already discovered earlier that light travelled at the constant speed of 300,000 km /sec, no matter the speed of its source.[1]

---

1   See the discoveries by the astronomers 0. Roemer (1676) and J. Bradley (1727), and the physicists A. Fizeau (1849) and J. Foucault (1850).

The phenomenon of electricity not only left its imprint on the scientific community, it also stimulated business activities. Electricity changed the way people lived. Generators were invented, dynamos suddenly became available in the market, and incandescent bulbs lighted the houses, streets, docks, and railways of the cities, whereas the telegraph connected London and Berlin. No wonder that the young Einstein, whose father was engaged, as a businessman, in the manufacture of electronic equipment, was fascinated by the properties of light waves. At age sixteen "he asked himself what would be the consequences of his being able to move with the speed of light. This question, innocent as it appears, brought him into conflicts and contradictions of enormous depth within the foundations of physics."[2] As a teenager he had already been engaged in a private study of the classics of physics, dissatisfied as he was with the rote learning at the *Luitpold* high school in Munich. The grades he received in high school were mediocre. He hated high school so much that, in his last year, he took an early entrance examination for the Swiss Polytechnic Institute in Zürich, but failed. Yet, after one year of intensive study in Aarau, Switzerland, he was able to successfully complete the same examination. It is at this Polytechnic Institute that he had the privilege of having the renowned mathematician Hermann Minkowski as his teacher. Once he graduated, Einstein took a menial job as a clerk in the Swiss Patent Office in Bern. This was enough for him to earn his living so that he could devote his spare time to the cherished thought-experiments that eventually would revolutionize the Newtonian notions of space and time.

The constancy of the speed of light would reveal a major problem to Einstein, for it basically exploded the framework of Newtonian physics. Indeed, within that framework the speed of light varied, depending on whether one measured it from a stationary or a moving platform. If assessed from a stationary frame of reference, the speed of light is measured at 300,000 km/sec, whereas assessed from a moving frame of reference (in the direction of the light source) it is measured at less than 300,000 km/sec: "Common sense would seem to tell us that if we were to travel very rapidly in some direction, then the speed of light in that direction ought to appear to us to be *reduced* to below c (because we are moving towards 'catching the light up' in that direction)."[3] The speed (v) with which we, when sitting in a space craft, rush towards the beam of

---

2   Jeremy Bernstein, *Einstein* (Glasgow: Fontana,1976), 39.

3   Roger Penrose, *The Emperor's New Mind: Concerning Computers, Minds, and the Laws of Physics* (Oxford: Oxford University Press, 1990), 191.

light ought, in this logic, to be subtracted from the speed of that beam of light (c): the resulting velocity W should be: "W = c − v."[4]

We have seen this procedure of subtraction and addition at work in Newton's description of the sailor walking on the deck with a mixture of absolute and relative motion. The sailor moves at a velocity of 10,001 units eastward: namely the velocity of walking on the deck (+1) added to the velocity of the rotating Earth, which also goes east (+10,010), minus the velocity of the ship sailing west (-10). Yet, such computations militate against the constant value of the speed of light in Maxwell's field equations. These field equations give the same results whether they are applied to a stationary frame of reference or to a frame of reference moving uniformly in a straight line. "It was through worrying about such matters that Einstein was led in 1905 [...] to the special theory of relativity."[5] He was not willing to give up the elegance of what is technically called the "equivalence principle," according to which "the physical laws [...] remain totally unchanged if we pass from a stationary to a moving frame of reference,"[6] an insight Galileo gained with his experiments.

Relativity of Time and Space

The speed of light remains invariant if one passes from one frame of reference to another frame of reference. Based on this insight, Einstein will challenge Newton's notions of absolute space and time. In his popular exposition of the theory of relativity, he gives the following examples to illustrate the non-absolute character of time and space.

In a first move, Einstein dismantles the classic notion of simultaneity. He begins with constructing two different frames of reference: a stationary "railway embankment" and a "very long train travelling along the rails with the constant velocity v and in the direction indicated in the figure below." He then asks the question as to whether two strokes of lightning, A and B, that from point M on the embankment are registered as simultaneous, will also appear as simultaneous for an observer in the train who finds himself at point M':

4   Albert Einstein, *Relativity. The Special and the General Relativity: A Popular Exposition*, trans. Robert Lawson (New York: Bonanza Books, 1961), 18.

5   Roger Penrose, *The Emperor's New Mind*, 191–92.

6   Ibid., 191.

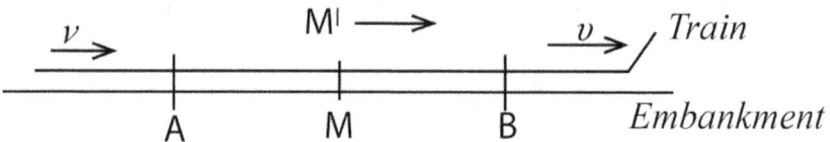

Figure 14. Einstein's "railway embankment"

Source: Albert Einstein, *Relativity: The Special and General Relativity*, trans. Robert W. Lawson
(New York: Henri Holt and Company, 1921), 30; Cornell University Library.
https//archive.org/details/cu31924011804774 (accessed September 22, 2014).

The answer to this question is in the negative. It is indeed obvious that given the speed of the train the observer sitting at M' will hear the thunderbolt coming from B earlier than that coming from A, in contrast to the stationary observer at M (on the embankment) who hears them simultaneously. Similarly, the strokes of lightning will not be seen simultaneously by the observer sitting at M'; yet, the time interval between both will be perceived as much shorter than the time interval of the thunderbolts, since light waves propagate themselves so much faster than sound waves.

> Observers who take the railway train as their reference-body must come to the conclusion that the lightning flash B took place earlier than the lightning flash A. We thus arrive at the important result: Events which are simultaneous with reference to the embankment are not simultaneous with respect to the train, and *vice versa* (relativity of simultaneity). Every reference-body (coordinate-system) has its own particular time; unless we are told the reference-body to which the statement of time refers, there is no meaning in a statement of the time of an event.

This insight revolutionizes the classic notion of time: "Before the advent of the theory of relativity it had always tacitly been assumed in physics that the statement of time had an absolute significance, i.e., that it is independent of the state of motion of the body of reference."[7] Yet, the above experiment has shown that time depends on the velocity of the body/frame of reference.

The same is true for the measurement of space. Take again a train traveling along the embankment with a velocity v. If one would measure *in the train* the distance between the middle of the first and of the

---

7   Albert Einstein, *Relativity*, 26–27.

twentieth carriage then it would turn out that the measurement of that same distance conducted *on the embankment* would be slightly different: "*A priori* it is by no means certain that this last measurement would supply us with the same measurement as the first. Thus the length of the train as measured from the embankment may be different from that obtained by measuring in the train itself."[8] Measured from within the train, the length of the train turns out to be shorter.

For Einstein, the specific values one gets for time and distance are clearly dependent on the condition of motion (that is: on the speed) of the frame of reference. Yet, in spite of that fact, the speed of light is always perceived as travelling at 300,000 km/sec: in all frames of reference with their varying speeds. The speed of light is constant and is not in the least affected by either the speed of the object emitting the light, *or by the speed of the observer perceiving it.* Let us note in passing that, strictly speaking, the embankment on which the measurements are made is also 'in motion' due to the rotation of the Earth.

Length Contraction and Time Dilatation

In order to corroborate the above insights Einstein has recourse to the Lorentz-transformations,[9] which describe how measurements of space and time by two observers moving at different velocities are related. On the basis of the Lorentz-transformations it can be demonstrated that rods in motion, which measure length, contract, and that clocks in motion, which measure time, go slower: "the rigid rod is shorter when in motion than when at rest, and the more quickly it is moving, the shorter is the rod." Similarly it must be said that "as a consequence of its motion the clock goes more slowly than at rest." For Einstein, the length contraction and time dilatation are the result of the fact that the speed of light is "a limiting velocity, which can neither be reached nor exceeded by any real body."[10] It looks as if the speed of light were really preventing measure-rods and clocks from coming up with results that would exceed the speed of light: so rods must be shortened and clocks slowed down the more they approach the speed of light.

---

8   Ibid., 29.

9   The Dutch physicist Hendrik Lorentz examined the conditions under which Maxwell's equations were invariant when transformed from the ether (the stationary medium in which the electromagnetic waves were believed to propagate) to a moving frame (the Earth).

10   lbert Einstein, *Relativity*, 35–37.

A great many of the effects of relativity are counter-intuitive. No wonder that the observers within a particular frame of reference are unaware of the changes in their own measuring instruments. Take two spacecrafts, each with an observer (a pilot) in them. The spacecrafts fly parallel to each other, each with a relative uniform motion and in the same direction. The point of view of each will be that the other's (moving) clock is ticking at a *slower* rate than their local clock and that the other's (moving) measure rod is *shorter* than their local measure rod. In other words, when looking into each other's cabin, the pilots will perceive that—in their neighbor's cabin—time is running slower and, what is still more remarkable, that their neighbor's spacecraft (at a great speed approaching the speed of light) is much shorter than it was on the ground. A spacecraft of 60 meters length at rest, for example, shrinks to 59 meters when reaching 10 % of the speed of light and to 30 meters at 86.5 % of the speed of light. What it really means to say that time is running slower in comparable circumstances is, in turn, not easy to imagine. It is illustrated, however, by the "twin paradox" of relativity:

> one twin brother remains on the earth, while the other makes a journey to a nearby star, travelling there and back at a great speed, approaching that of light. Upon his return, it is found that the two twins aged differently, the traveler finding himself still youthful, while the stay-at-home brother is an old man.[11]

Equivalence of Mass and Energy

Einstein regards the equivalence of mass and energy as one of the most important results of the special theory of relativity: a body's mass can be converted into energy, and likewise energy can be converted into mass. This convertibility presupposes that mass is not constant: A body's mass depends on the speed of its frame of reference, in the same manner as duration and length depend on the speed of their frame of reference.

> A moving electron, or any massive object, becomes more massive when it is in motion with respect to an observer than when it is at rest with respect to the same observer. In particular, as the speed of the object approaches the speed of light its mass becomes infinite.[12]

---

11   Roger Penrose, *The Emperor's New Mind*, 197.

12   Jeremy Bernstein, *Einstein*, 80.

Here again, the speed of light is a limit concept. It can never be reached by a material object, for

> as an object approaches the speed of light, its mass rises ever more quickly, so it takes more and more energy to speed it up further. It can in fact never reach the speed of light, because by then its mass would have become infinite, and by the equivalence of mass and energy, it would have taken an infinite amount of energy to get it there.[13]

For Einstein, there exists a strong interrelation between mass and energy. To express this interrelation mathematically, Einstein elaborated the simple and by now famous equation: $E = mc^2$, where E denotes the energy of the object, m the mass, and c the speed of light. Energy equals mass multiplied by the square of the speed of light.

This equation also suggests the possibility of transforming mass into energy and vice versa. Its most spectacular effect is the transformation of mass into energy which Einstein brought up in one of his 1905 papers entitled *Does the Inertia of a Body Depend upon Its Energy Content?*: "If a body gives off the energy E in the form of radiation its mass diminishes by E/c^2."[14] Conversely, the radiation that is set free has the power of the diminished mass—a power that, as we now know, may cause a massive nuclear explosion, as this happened with the dropping of two atomic bombs in Hiroshima and Nagasaki with an estimated death toll of 300,000 people. "The fact that mass is equivalent to energy means that, in a sense, matter is 'locked up' energy. If some way can be found to unlock it, matter will disappear amid a burst of energy. Conversely, if enough energy is somehow concentrated, matter will appear."[15]

Four-Dimensional Space-Time

Einstein's picture of the universe presupposes a special geometry that departs from Euclidean geometry. This special geometry was developed by Hermann Minkowski, three years after Einstein launched his special theory of relativity:

> Minkowski had been one of Einstein's teachers at the Zürich Polytechnic Institute. His fundamental new idea was that space and time had to be considered together

---

13   Stephen Hawking, *A Brief History of Time* (London: Bantam Books, 1989), 21.

14   Quoted in Jeremy Bernstein, *Einstein*,84.

15   Paul Davis, *God and the New Physics* (London: Penguin Books, 1983), 26.

as a single entity: a *four-dimensional space-time*. In 1908, Minkowski announced, in a famous lecture at the university of Göttingen: Henceforth space by itself and time by itself, are doomed to fade away into mere shadows, and only a kind of union of the two will preserve an independent reality.[16]

The best way to get an idea of Minkowski's geometry is to examine what he reveals about the light cone. This light-cone is, in fact, a double-

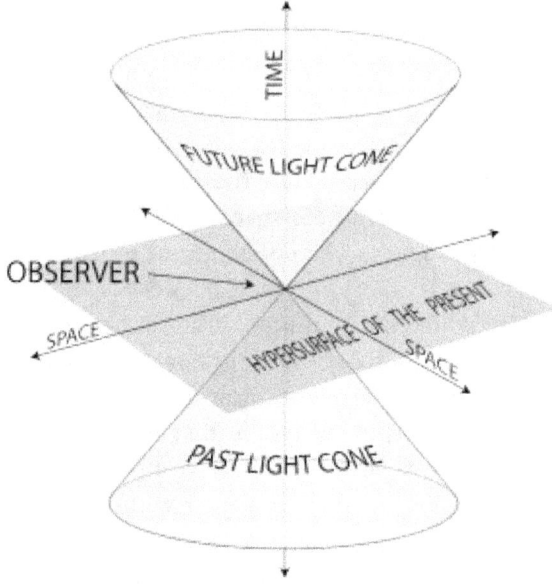

Figure 15. Light cone

Source: Wikipedia (http://en.wikipedia.org/wiki/Light_cone), accessed September 22, 2014.

cone, centered at each event in space-time. The upper-cone (the future light cone) represents the history of a light-flash emitted at that event. The lower cone (past light cone) represents all directions from which light flashes can be received at that event. Future and past light cones move outward at the speed of light.

Let us for purposes of explanation focus on the future light cone. This will allow us to grasp what Minkowski has in mind: to give a geometrical presentation of the basics of special relativity. That which was called "the event" at which the future light cone is centered can also be termed a point of "space-time origin."[17] In this scenario it would look as if a portion of space-time pops up and, as in an explosion, gives rise to particles and to the future light cone that moves outward at the speed of light. The future light cone is the area in which particles enact their history. Since the particles persist in time they are represented as lines: the "world-lines." It is these world-lines that are going to travel within the light cone. The world-lines of particles that move uniformly are straight, whereas those of accelerating particles are curved.

> Now, one of the features of relativity theory is that it is impossible for a material particle to travel faster than light. All the material particles coming from the explosion must lag behind the light. This means, in space-time terms, that the world-lines of all the material particles emitted in the explosion must lie inside the light cone.[18]

For photons, which are massless particles, this is different. Photons travel at the speed of light. So it is quite natural that the world-line of a photon is "always along the light cone at each point, whereas the history [the world-line] of any material particle must always be inside the light cone at each point."[19]

A further characteristic of special relativity is that the more the material particles' world-lines dash further into the future swelling light cone—which moves outward at the speed of light—the more the clocks attached to them go slower. This again indicates that no material particle can travel faster than light.

The geometry of the light cones with growing circle makes it clear that (a) no signal of a cosmic event can reach us that would be traveling faster than light, and (b) that only in places that lie within this growing light cone the emitted signal can be registered.

> For example, if the sun were to cease to shine at this very moment, it would not affect things on Earth at the present time because they would be in the elsewhere of the event when the sun went out. We would know about it only after eight minutes, the time it takes light to reach us from the sun. Only then would events on Earth lie in the light cone of the event at which the sun went out. Similarly, we do not know what is happening at the moment farther away in the universe:

---

17   Roger Penrose, *The Emperor's New Mind,* 193.
18   Ibid., 194.
19   Ibid., 195.

the light that we see from distant galaxies left them millions of years ago, and in the case of the most distant object that we have seen, the light left some eight thousand million years ago. Thus, when we look at the universe, we are seeing it as it was in the past.[20]

For Einstein, the four-dimensional space-time revolutionizes our notion of time. In space-time there is no longer an independent, absolute notion of time with universal value: "'now' loses for the spatially extended world its objective meaning."[21] It would be nonsensical to regard our 'now moment' of observation as coinciding with the proper time sequence of faraway events in the universe. If one looks, for instance, at light emitted by a star, one can be sure that its very source is no longer at the place where one perceives it to be now. There are perhaps several (million) light-years between the light emission and our observation of it, and during that interval, the star changed its position.

General Theory of Relativity

The special theory of relativity is restricted to the study of *uniform* straight line motions. General relativity, however, includes the study of *non-uniform* motions, such as acceleration, retardation, and rotation. Yet, such an extension is not simple; it will necessitate, as we will see, the elaboration of a geometrical theory in which space is curved, an approach that definitely goes beyond Minkowski's geometry of four-dimensional space-time.

Reduced to its essence, special relativity maintains that the same physical laws (including the propagation of light) hold with reference to a framework when at rest (the embankment) or in *uniform* motion (the railway-carriage, to use again this illustration). Now, one's instinct for generalization would be tempted to extend this equivalence to *all sorts* of motions. But, at a first glance, there is little hope of the success of such an attempt.

To give us an idea of the fact that such a generalization is not immediately plausible, Einstein constructs a scenario in which it becomes evident that a "retardation" disrupts the equivalence of a state of rest and a state of uniform straight line motion:

---

20   Stephen Hawking, *A Brief History of Time*, 30.

21   Albert Einstein, *Relativity*, 149.

(a) Let us imagine ourselves transferred to our old friend the railway carriage, which is travelling at a uniform rate. As long as it is moving uniformly, the occupant of the carriage is not aware of its motion, and it is for this reason that he can without reluctance interpret the facts of the case as indicating that the carriage is at rest, but the embankment in motion. [...] (b) If, however, the constant motion of the carriage is interrupted, or changed into a non-uniform motion, as for instance by a powerful application of the brakes, then the occupant of the carriage experiences a correspondingly powerful jerk forwards. The retarded motion is manifested in the mechanical behavior of bodies relative to the person in the railway carriage. The mechanical behavior is different from that of the case previously considered, and for this reason it would appear to be impossible that the same mechanical laws hold relatively to the non-uniformly moving carriage, as hold with reference to a carriage when at rest or in uniform motion.[22]

Indeed, because of the change into a non-uniform motion, the occupant of the carriage no longer feels at rest. The fact that he experiences a powerful jerk forward suddenly interrupts his state of rest, and so the "equivalence" between a state of rest and a uniform straight line motion no longer holds. As we will see in more detail below, the trouble maker is apparently gravity, which was not included in the special theory of relativity. Yet, if one takes gravity as a new absolute, then equivalence between a state of rest and a state of non-uniform motion (acceleration, retardation) can be established. That is what Einstein undertakes in his general theory of relativity.

Equality of Gravitational and Inertial Mass

In order to make plausible the new equivalence, Einstein shows, in a first move, that "the gravitational mass is equal to its inertial mass."[23] The two sorts of mass already figured in Newton's theory of motion: Inertial mass m measures the body's acceleration as a response to a given force F—see Newton's second law of motion $F = ma$—whereas gravitational mass measures the gravitational attraction two bodies exert on each other.

Einstein, first of all, points to an inconsistency in Newton's theory of gravity. According to this theory gravity works at a distance through the void, and its effect is instantaneously felt: for Newton, the gravitational

---

22   Ibid., 61–62 (translation slightly modified).
23   Ibid., 65.

attraction of the Sun is immediately felt by the Earth. For Einstein, this is improbable for two reasons. It would, first of all, presuppose a communication faster than the speed of light. And second, it ignores the fact that gravity is propagated by gravitational fields.

When reflecting on gravity, Einstein from the outset links it to gravitational fields. For him, the answer to the question "If we pick up a stone and then let it go, why does it fall to the ground?" is simply: "because it is attracted by the gravitational field of the earth." The procedure is similar to a magnetic field that attracts a piece of iron. Yet,

> in contrast to electric and magnetic fields, the gravitational field exhibits a most remarkable property, which is of fundamental importance for what follows. Bodies which are moving under the sole influence of a gravitational field receive an acceleration, *which does not in the least depend either on the material or on the physical state of the body*. For instance, a piece of lead and a piece of wood fall in exactly the same manner in a gravitational field *(in vacuo)*, when they start off from rest or with the same initial velocity,[24]

a phenomenon Galileo had already discovered with his experiment of bodies rolling down a smooth slope.

It is because this acceleration happens *at the same rate* for all falling bodies, no matter what their weight, that an equality of gravitational and inertial mass can be postulated: "If, as we find from experience, the acceleration is to be independent of the nature and the condition of the body and always the same for a given gravitational field, then the ratio of the gravitational and the inertial mass must likewise be the same for all bodies."[25] This insight will lead Einstein to the elaboration of the equivalence of "acceleration due to gravity" and a "state of rest."

Equivalence of "Acceleration Due to Gravity" and "State of Rest"

> Einstein's first paper on this subject was published in 1907. 13 years later he commented that while writing it a thought came to his mind, which he called "the happiest thought of my life": "*The gravitational field has only a relative existence...*because for an observer freely falling from the roof of a house—*at least in his immediate surroundings*—there exists no gravitational field."[26]

---

24   Ibid., 64.

25   Ibid., 65.

26   Joseph Wudka, *Space-Time, Relativity, and Cosmology* (Cambridge: Cambridge University Press, 2006), 200.

For somebody finding himself in a free fall movement, this fall is experienced as a state of rest as long as he shuts his eyes (or confines his observations to his immediate vicinity). This implies that a state of rest is indistinguishable from acceleration due to gravity.

In his popular exposition of relativity, Einstein brings this insight home to his readers with the help of what has come to be known as the "Einstein elevator." "This is a closed box sitting in space somewhere which can be tugged , say, 'up' by someone outside pulling on a rope attached to the roof, with a constant force," [27] so that the box is constantly being pulled upward. In the first moments the occupants of the elevator will feel pressed "down" toward the floor, but after a while they are unable to distinguish whether they are in a state of acceleration or at rest. Einstein uses this example to demonstrate the equivalence of acceleration due to gravity and a state of rest as experienced by the occupants of the elevator. He does this in various steps. (a) One of the occupants drops a ball and sees that it goes right down to the floor of the elevator in an *accelerated relative motion*; the same is true for objects of different weight. Relying on his knowledge of gravitational fields, he infers from it that the box must feel the effect of a gravitational field, (b) Realizing this, he begins to be puzzled as to why the box does not fall into the gravitational field. At that moment and just by chance, he looks up at the ceiling of the box and perceives there a hook on which the rope is attached that pulls the box upward, and comes to the conclusion that the box must be suspended *at rest* in a gravitational field. "Ought we to smile at the man," Einstein asks, "and say that he errs in his conclusion? I do not believe we ought to if we wish to remain consistent; we must rather admit that his mode of grasping the situation violates neither reason nor known mechanical laws. Even though it is being accelerated [...] we can nevertheless regard the box as being 'at rest.'"[28]

Einstein calls attention to the fact that this double interpretation— being in a state of accelerated motion and being at rest—is only possible because of the special property of the gravitational field of giving all the bodies, whatever their mass, the same rate of acceleration. If this property did not exist, the man in the accelerated box "would not be able to interpret the behavior of the bodies around him [i.e., of the objects he dropped] on the supposition of a gravitational field, and he would not be

27   Jeremy Bernstein, *Einstein*, 101.

28   Albert Einstein, *Relativity*, 67–68.

justified on the grounds of experience in supposing his reference-body to be at rest."[29]

The equivalence of "acceleration" and "state of rest" resolves also the problem that Einstein initially brought up, namely that 'being at rest' would be incompatible with an accelerated motion. He comes back to the story of the observer in the railway carriage who experienced a jerk forwards as a result of the application of the brakes, and concluded from this that the carriage was suddenly accelerated or retarded. Yet, Einstein writes,

> the observer is compelled by nobody to refer this jerk to a "real" acceleration (retardation) of the carriage. He might also interpret his experience thus: "My body of reference (the carriage) remains permanently at rest. With reference to it, however, there exists (during the period of application of the brakes) a gravitational field which is directed forwards [...]. Under the influence of this field, the embankment together with the earth moves non-uniformly,"[30]

and this accelerated motion of the earth (and of the embankment) causes the jerk that was felt in "the carriage at rest."

With these illustrations Einstein made it clear that in the general theory of relativity, with its focus on acceleration, the basic equivalence principle established by Galileo also holds, "namely that the physical laws remain unchanged if we pass from a stationary to a moving frame of reference."[31] This means in the concrete that "the effects of a uniform constant acceleration on an observer or on his measuring instruments are indistinguishable from—that is, equivalent to—the observer's being at rest but acted on by a uniform field of gravitation."[32] Yet, whereas in special relativity such an equivalence was reached on the basis of the constancy of the speed of light (which "relativizes" any other gradation of regular motion), in the case of general relativity this equivalence obtains on the basis of the constant effect of the uniform gravitational field. This difference in background explanation will create a new problem, as will become evident in the "bending of light."

---

29   Ibid., 68.

30   Ibid., 70.

31   Roger Penrose, *The Emperor's New Mind*, 191.

32   Jeremy Bernstein, *Einstein*, 100.

Special Relativity as a Limiting Case of General Relativity:
the Bending of Light

The special theory of relativity started from the assumption that light—the new absolute standard—is rectilinearly propagated. This assumption can no longer be upheld in the context of general relativity, for in it the power of the gravitational field is such that it makes the straight rays of light curve. This phenomenon is known as the 'bending of light'. This is illustrated as follows:

> Let us imagine the following situation: an elevator is attached to its rope and being pulled upward with a constant force and hence a uniform acceleration. We are stationed outside the elevator in the "rest frame" with respect to which the elevator is accelerating. We now fire a beam of light from our rest frame in such a way that the light enters the elevator—by a small window, if you will—on a trajectory that is initially parallel to the elevator floor. What we will observe happening is the floor of the elevator accelerating upward toward the light beam. To us in our rest frame the light ray follows a straight line, while to the people in the elevator the light beam will appear to have been bent down toward the floor in an arc. If they do not "know" that they are being pulled up, they can, according to the principle of equivalence, conclude that there is a uniform grav-itational field in their region of space which is bending the light downward in a curved path.[33]

This bending of light is, again, very strange, since according to Newtonian physics, gravity only acts on material objects with mass, whereas light is apparently massless. If one knows, however, about the formula $E = mc^2$, which posits the equivalence of energy and mass, then this strangeness recedes. A beam of light transports energy; on this basis it can be bent down by a gravitational field: "*Rays of light are propagated curvilinearly in gravitational fields.*"[34]

This finding, however, leads to an apparent conflict with the special theory of relativity which rests on the assumption of the *constancy* of the speed of light. Yet, "a curvature of rays of light can only take place when the velocity of propagation of light varies with position." The more a ray of light curves, the slower it goes, which means that it will travel *slower* than 300,000 km/sec. So, "the special theory of relativity cannot claim an unlimited domain of validity; its results hold only so long as we are able

---

33   Ibid., 101.

34   Albert Einstein, *Relativity*, 75.

to disregard the influences of gravitational fields on the phenomena (e.g. of light)."[35] On the other hand, what the special theory of relativity tells us about the constancy of the speed of light has a tremendous *research value*: it allows us to realize the dramatic changes brought about by the presence of gravitational fields (such as the bending of light, and the slowing down of its velocity of propagation). "In the example of the transmission of light just dealt with, we have seen that the general theory of relativity enables us to derive theoretically the influence of a gravitational field on the course of natural processes, *the laws of which are already known when a gravitational field is absent*." It is thanks to the insights gained from the special theory of relativity that the general theory of relativity could develop. The general theory of relativity is a comprehensive theory, in which the special theory of relativity "lives on as a limiting case."[36]

The bending of light has been experimentally tested. According to Einstein's calculations a light ray passing closely by the Sun's surface would undergo a deflection of 1.7 seconds of arc under the influence of the Sun's gravitational field. Such a deflection, however, can only be observed during a solar eclipse. In 1919 two expeditions equipped by the Joint Committee of the Royal Society and the Royal Astronomical Society in London, one in Sobral in northern Brazil, and the other on the Isle of Principe in the gulf of Guinea off West Africa, succeeded in making photographs of the solar eclipse on May 29. The results of their measurements confirmed Einstein's prediction of the light ray's deflection.

Warping of Space and Slowing Down of Time

Given the special role played by gravitational fields in the general theory of relativity, Einstein needed to elaborate a new concept of geometry: one that, more drastically than Minkowski's geometry of four-dimensional space-time, departs from Euclidean geometry. The system of straight lines and coordinate time used in Euclidean geometry is no longer workable for accelerating or rotating frames of reference.

In a first move Einstein shows that Euclidean geometry no longer holds in the case of a rotating disc. (a) If one measures the circumference of a rotating disc and divides it by the diameter of the disc one no longer gets the celebrated number known as 'pi' (3. 1415927...), as to be expected in the Euclidean universe, but a number larger than 'pi': "This proves,"

---

35   Ibid., 76.
36   Ibid., 77 (italics mine).

Einstein remarks, "that the propositions of Euclidean geometry cannot hold exactly on the rotating disc, nor in general in a gravitational field. [...] Hence the idea of a straight line also loses its meaning."[37] (b) If one places two identical clocks, one on the circumference of a rotating disc and one close to the rotating disc's center, and compares the registered times, it turns out that the clock on the circumference runs slower than the central clock. From this state of affairs Einstein concludes that "it is not possible to obtain a reasonable definition of time[...]. On our circular disc, or to make the case more general, in every gravitational field, a clock will go more quickly or less quickly, according to the position in which the clock is situated (at rest)."[38]

When both the idea of a straight line and of coordinate time become meaningless in the context of accelerated motions, then a new *non-Euclidean geometry* will be required in which space curves and time slows down with increasing acceleration, or what boils down to the same: with the growing effect of a gravitational field. Such geometry had already been developed in the 19th century by Carl Friedrich Gauss (1777–1855), and Bernhard Riemann (1826–1866). Following the lead of Gauss, Einstein set out to construct a non-Euclidean continuum in which Gaussian coordinates play an important role. Euclidean continuum is a space in which one can reach any point by first reaching a nearer point, and in which it is taken for granted that the points are linked to each other by straight lines from which adjacent squares can be formed. A non-Euclidean continuum, on the contrary, will be formed by curved, arbitrarily chosen intersecting lines, termed Gaussian coordinates, and such a system can be extended to a number of dimensions.

When working with Gaussian coordinates one no longer assumes that distances can be measured with rigid rods and time sequences with regular clocks. Such measurements only hold for Euclidean, "flat" geometry; it is only in this "flat" geometry that the angle sum of a triangle is 180 degrees. Besides this, there are geometries for curved surfaces. In the geometry developed by Riemann, for instance, all triangles have angle sums *greater* than 180 degrees. One can visualize this as the geometry with "positive" curvature, like a sphere. In the "hyperbolic" geometry of Gauss, on the contrary, "all triangles have angle sums *less* than 180

---

37   Ibid., 82.
38   Ibid., 81.

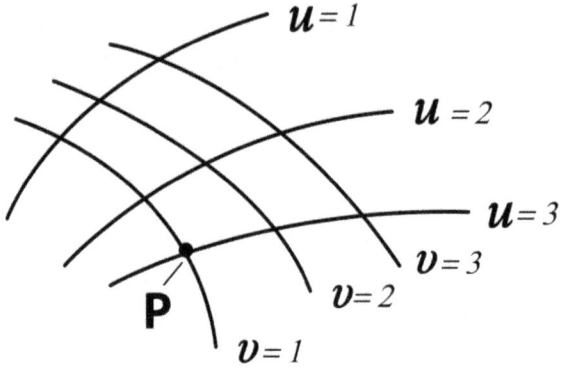

Figure 16- Gaussian coordinates

Source: Albert Einstein, *Relativity: The Special and the General Theory*, trans. Robert W. Lawson (New York: Henri Holt and Company, 1921), 103; Cornell University Library https//archive.org/details/cu31924011804774 (accessed September 22, 2014).

degrees. One can visualize this as the geometry of 'negative' curvature shaped something like a funnel,"[39] or a saddle.

It is the geometry of "negative curvature" that challenged Einstein to reformulate his equivalence principle. This principle now reads as follows: "*All Gaussian coordinate systems are essentially equivalent for the formulation of the general laws of nature*,"[40] i.e., they are all equivalent for the formulation of the effect of gravitational fields on the warping of space-time. The Gaussian coordinate systems are very flexible: they move in any way whatsoever and may suffer all possible alterations during their motion. Nonetheless, all these flexible coordinate systems "can be used as reference-bodies with equal right and equal success in the formulation of the general laws of nature."[41]

Einstein's Post-Newtonian Cosmology

Einstein goes beyond Newton mainly in three areas: his new interpretation of gravitation; his accurate calculation of the recession of the orbit of Mercury; and his prediction of a non-static universe.

1. The concept of gravitation. Newton avowed that he did not exactly know what gravitation was, but that, in spite of this, he was able to

---

39   Jeremy Bernstein, *Einstein*, 122.

40   Albert Einstein, *Relativity*, 97.

41   Ibid., 99.

formulate the equation that ruled its effects: $F = Mm/r^2$: the force that attracts two bodies to each other equals the product of their respective masses (Mm), whereas the pull of this force diminishes with the square of the distance (r). Based on this equation, he could predict, for example, the velocity with which each of the planets orbit the Sun on their elliptical trajectory. Newton's equations are still essentially valid and continue to be used for rough calculations, but Einstein's approach is more refined. "In Einstein's new approach the equations of Newtonian gravitation are replaced by a new set of equations that retain their form *in all possible coordinate systems*, both accelerated and uniformly moving."[42]

The fact that they retain this form in all possible coordinate systems is intrinsically connected with the phenomenon of gravitational fields.

Figure 17. Einstein's warped space

Source: NASA Science (http://nasa.gov/science-news/science-at-nasa/2005/28mar_gamma/), accessed September 22, 2014.

Einstein made the revolutionary suggestion that gravity is not a force like other forces, but is a consequence of the fact that space-time is not flat, as had been previously assumed, but curved, or "warped," by the distribution of mass and energy in it.[43]

---

42   Jeremy Bernstein, *Einstein*, 113.

43   Stephen Hawking, *A Brief History of Time*, 32.

Wherever in the universe there is a significant amassment of mass and energy, gravitational fields make their appearance, and this leads to a "warping" of space (and a "slowing down" of time). If one takes our Sun, for example, then it is evident that its mass and energy make the surrounding space adopt a hollow curvature.

This concept of "warped" space would have been completely incomprehensible to Newton. For him, the force of gravity pierced through the void and used its power to maintain the planets in their orbits. Not so, for Einstein, for whom the planets do not follow circular or elliptical paths on a flat plane, but rather follow a straight line within a curved region of space.

> Bodies like the earth are not made to move on curved orbits by a force called gravity; instead, they follow the nearest thing to a straight path in a curved space, which is called a geodesic. [...] This is rather like watching an airplane flying over hilly ground. Although it follows a straight line in three-dimensional space, its shadow follows a curved path on the two-dimensional ground.[44]

The curvature of space caused by the gravitational field generated by the Sun's mass/energy is, so to speak, the natural milieu in which the planets move in quasi-straight lines: the nearer they are to the Sun the faster will they complete their journey through that part of curved space-time.

2. The recession of the orbit of Mercury. In Newton's universe each planet orbits the Sun along an elliptic path; the Sun is not located in the center of the ellipse but slightly off-center, so that two areas can be distinguished: the perihelion or that part of the ellipse that is closest to the Sun, and the aphelion or that part that is farthest away from the Sun. Newton had already observed that after one orbit the planet did not return exactly to the same place: after each journey around the Sun the planet's aphelion slightly receded from its previous position. Newton realized that this recession was due not only to the Sun's attraction but also to the attraction that the other planets exerted on the planet in question.

This recession was the most visible in the case of Mercury, the planet closest to the Sun: "Each time the planet went once around the Sun it

---

44   Ibid., 32–33.

returned to a different point in space, so that the orbit, if viewed long enough, would look something like the petals of a flower."[45] This was an occasion for Newton to calculate the recession of that planet. He ascertained that it would take Mercury 64 years and 286 days to return to its original position. With the help of his field equations, Einstein made the same calculations and discovered that Mercury's return to the original position would take 182 days less. This data was confirmed by subsequent astronomical observations.

3. Prediction of a non-static universe. Whereas Newton still retained a picture of the world in which the solar system was in the center of the universe, Einstein laid down the foundation for a universe that has no proper center. When elaborating in 1917 the equations representing the whole of the universe he started from the assumptions that at large scales the universe is homogeneous (uniform in composition throughout) and isotropic (uniform when measured from different directions). These assumptions had far-reaching consequences; they, in fact, annulled the central position previously ascribed to our solar system. In Stephen Hawking's words:

> Now at first, all this evidence that the universe looks the same whichever direction we look in might seem to suggest that there is something special about our place in the universe [...] There is, however, an alternate explanation: the universe might look the same in every direction as seen from any other galaxy, too.[...] It would be most remarkable if the universe looked the same in every direction around us, but not around other points in the universe![46]

But there is more to it. Einstein's equations predicted a universe that was not static, but caught up in a process of continuous formation: his equations represented universes that were either expanding or contracting; there were no solutions representing a static universe."[47] In 1917, when Einstein was elaborating his equations, the telescopes were not yet strong enough to explore parts of the universe outside our own galaxy. So, since the predicted expansion was not observable within our galaxy, Einstein felt compelled to "modify his equations in such a way as to allow solutions

45   Jeremy Bernstein, *Einstein*, 126.

46   Stephen Hawking, *A Brief History of Time*, 45.

47   Joseph Wudka, *Space-Time, Relativity and Cosmology*, 230.

describing a static universe, while still preserving the geometrical character of the theory."[48] The next generation of scientists would demonstrate that such a modification was not necessary and that what Einstein's original equations predicted—an expanding universe—was the true picture of the world.

---

48   Ibid., 232.

CHAPTER 9

Quantum Physics

With the study of quantum dynamics, we move from the world of space-crafts and rushing galaxies to that of subatomic elements: "Quantum theory arose from an attempt to explain phenomena that lay beyond the scope of conventional classical physics. A central failure of classical mechanics was its inability to account for the structure of atoms."[1]

Exploration of the Atom

Until the end of the 19th century it was thought that the atom was the smallest indivisible unit of matter. This view was challenged by J.J. Thompson's discovery in 1897 of the existence of electrons within the atom. While doing experiments with cathode rays Thompson discovered that these rays were composed of very small, negatively charged particles, which he called "electrons." On this basis, he imagined the atom to consist of a number of electrons swimming in a sea of positive electric charge, so that the whole looked something like a raisin pudding. In a further development, Ernest Rutherford in 1911 identified the atom's nucleus. He did so by bombarding an atom with heavy alpha particles, and from the manner in which the incident particles were deflected from the atom he deduced "that the bulk of the mass in the atom was concentrated in a tiny 'nucleus' of positive electric charge, with the electrons of negative charge distributed around it at assorted distances."[2] He termed the massive particles that

1   Jonathan Halliwell, "Quantum Cosmology and the Creation of the Universe," in *Scientific American,* December 7, 1991.

2   John Barrow, *The World Within the World* (Oxford/New York: Oxford University Press, 1990), 171.

composed the nucleus "protons," and also reckoned with the existence, in the nucleus, of hypothetical particles of similar mass to the proton but possessing no electric charge. These particles, the neutrons, were discovered by James Chadwick in 1932.

Rutherford's model of the atom was that of a solar system in miniature. But whereas in Newton's concept the planets orbited the sun under the impact of gravity, in Rutherford's model the electrons rotate around their nucleus on the basis of electrostatic forces. Unlike the solar system in which all the planets orbit the Sun on the same horizontal plane, each of the orbiting electrons has its own plane of circulation: horizontal, vertical, and various intermediate directions. This view, however, proved to be too simplistic. Indeed, in his study of electromagnetism, Maxwell had shown that "any electrically charged matter that would move non-uniformly must emit radiation, thereby decreasing its own energy."[3] So, as the radiating energy of a revolving electron continuously decreased, the electron should follow a spiral path and fall into the nucleus. This atomic model could, thus, not account for the stability of the atom.

In 1913 the Danish physicist Niels Bohr (1885–1962) who, as a young doctoral student, had worked in association with Rutherford in Great Britain, offered a solution to the problem by integrating Max Planck's new insight into the packet character of light radiation into the study of the subatomic particles.

In the 1890s Max Planck (1858–1947) was engaged in the study of thermal radiation, i.e., of the light radiation emitted from the surface of an object due to the object's temperature. The ideal thermal emitter is known as a black body, and the radiation it emits, when heated, is called black body radiation. When Planck began his study of black body radiation, it was already known that the amount of radiation did not depend on the material of the heated object but only on the latter's temperature. Yet, it was still unclear as to how the frequency of radiation related to the temperature and hence to the total energy of the black body: "In 1900 Max Planck suggested that light, X rays, and other waves could not be emitted at an arbitrary rate, but only in certain packets, called quanta. Moreover, each quantum had a certain amount of energy that was greater the higher the frequency of the waves."[4] This lead to the equation that would make

---

3   Mendel Sachs, *Einstein versus Bohr: The Continuing Controversies in Physics* (La Salle, Illinois: Open Court, 1988), 71.

4   Stephen Hawking, *A Brief History of Time. From the Big Bang to Black Holes* (London: Bantam Books, 1988 [1989]), 58.

him famous: $E = hv$; E stands for the total energy of the light source, v is the frequency of the light radiation, and h is a mathematical constant, called "Planck's constant." Planck's constant h expresses the packet character of the emitted radiation. The radiating energy of a system cannot take on any value. Rather, it can take on only particular discrete values. It is "as if nature allowed one to drink a whole pint of beer or no beer at all, but nothing in between."[5]

This principle sets the tone for Niels Bohr's atomic model and the restrictions it imposed: orbiting electrons cannot be located at *any* distance from the nucleus; they can only travel in special orbits: at a certain discrete set of distances from the nucleus with specific energies.

> Bohr's rules required that the *angular momentum* of electrons in orbit about the nucleus can occur only in integer multiples of $h/2\pi$,[6] for which Dirac introduced the convenient symbol $\hbar$ (pronounced h bar); that is: $\hbar = h/2\pi$. Thus the only allowed values of angular momentum (about any one axis) are 0, $\hbar$, $2\hbar$, $3\hbar$, $4\hbar$ etc.[7]

Bohr was apparently impressed by the way in which Planck "quantized" the light radiation with the help of his constant h: only discrete energy packets (quanta) can be emitted from the black body. So, too, Bohr emphasized that only specific distances from the nucleus qualify as orbits for the rotating electrons. Only at those distances is it possible for electrons to develop the exact energy and velocity that permanently keeps them on their orbit.

In a new move Bohr elaborated the shell-structure of the atom. The number of electrons in an atom is determined by the number of protons in its nucleus: A nitrogen atom, for example, has a nucleus made up of seven protons and seven neutrons; it is surrounded by seven electrons. For Bohr, the number of electrons an atom possesses is spread over various shells or energy levels. Bohr admitted of seven successive shells or energy levels, those being closest to the nucleus (but at a considerable distance from it) having the lowest energy level. Each shell has an upper limit to the number of electrons it can accommodate: the first shell is complete with two electrons (resulting in a helium atom), the second is complete with

---

5   Brian Greene, *The Elegant Universe* (New York: Vintage Books, 2000), 94.

6   Recall: π is 3. 1415927... "Pi" obtains if one divides the circumference of a circle by its diameter.

7   Roger Penrose, *The Emperor's New Mind: Concerning Computers, Minds, and the Laws of Physics* (Oxford: Oxford University Press, 1990), 231.

Bohr atomic model of a nitrogen atom

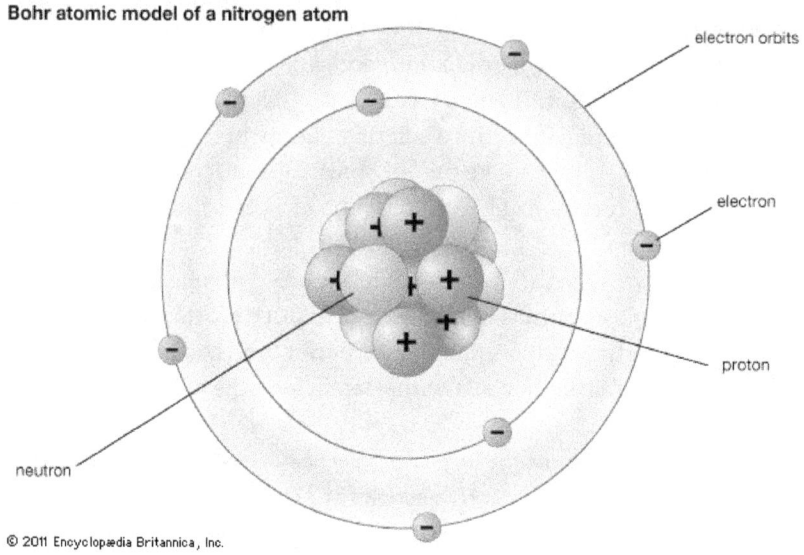

© 2011 Encyclopædia Britannica, Inc.

Figure 18. Bohr atomic model

Source: *Encyclopaedia Britannica* (http:Britannica.com/EBchecked/topic/71700/Bohr-atomic-model), accessed September 23, 2014.

eight electrons, and the other successive shells have still larger accommodation capacities. As a rule, a lower level must be totally filled before a higher shell can begin harboring electrons. Take the nitrogen atom with its seven electrons. Two of these electrons are in the first shell, whereas the remaining five are in the second shell. This rule of repartition holds for the first 18 elements in the periodic table of Mendeljev, but starting from the 19th element (the atom with 19 electrons) certain levels can be left incomplete. It is important to notice that the "last" electrons—those in the outer shell—determine the chemical properties of an atom. This is, for example, the case with uranium which has 92 electrons.

There is still one important element to be mentioned: the electrons can gain or lose energy by jumping from one allowed orbit to another, absorbing or emitting an electromagnetic radiation with a frequency v determined by the energy difference of the two levels. This is technically called a quantum leap. When a nitrogen atom, for instance, is bombarded with photons (i.e., quanta of electromagnetic radiation), one of its electrons may absorb the photons' energy and jump from its original energy level up to a higher energy level. This jump leaves an empty space in the lower energy level, which renders the atom unstable: the atom is then in an excited state. Yet, since the electrons always seek to get back to their most stable state in the atom, namely their ground state, they will eventually relinquish the newly acquired energy: the excited electron will drop back to its ground state by emitting a photon. It is this phenomenon of "ejection" that Einstein came across with his study of the photoelectric effect, although in that case an electron was knocked out of the atom.

The Copenhagen Interpretation

In the second half of the 1920s, the Copenhagen school became renowned for its method of statistical measurement. This method laid bare some unexpected features: the phenomena in the quantum world seemed to behave in a random, a-causal fashion, so that they ought to be described "in terms of a new probability calculus, called 'quantum mechanics'"[8] More particularly, the subatomic particles sometimes behaved as wave undulations and sometimes as packets of quanta (the so called "wave-particle duality"). The fact that light (photon) has both wave-like and particle-like properties was still easy to understand, but there were strong indications that matter particles, too, such as electrons, also had wave-like properties.[9]

The wave character of particles is clearly shown in the double-slit experiment that was originally performed by Thomas Young in 1801:

> Consider a partition with two parallel narrow slits in it. On one side of the partition one places a source of light of a particular color (that is, of a particular wavelength). Most of the light will hit the partition, but a small amount will go through the slits. Now suppose one places a screen on the far side of the partition from the light. Any point on the screen will receive waves from the two slits. However, in general, the distance the light has to travel from the source to the screen via the two slits will be different. This will mean that the waves from the slits will not be in phase with each other when they arrive at the screen: in some places the waves will cancel each other out, and in others they will reinforce each other. The result is a characteristic pattern of light and dark fringes.[10]

The light and dark fringes are the result of interference of waves: waves whose crests and troughs run parallel fortify each other, whereas waves whose crests and troughs do not run parallel cancel each other. Yet, other experiments reveal the weird character of the world of the quanta. If one replaces the source of light (the bulb) with a bunch of electrons, the result will be really mind-boggling. In that case electrons sent one after

---

8   Mendel Sachs, *Einstein versus Bohr*, 82.

9   This extension had been proposed by the French physicist Louis de Broglie in 1924.

10   Stephen Hawking, *A Brief History of Time*, 63.

the other to the partition with the two slits would behave as if *one single electron* went through *both of the slits* at once so as to produce the light and dark fringes, typical of interference. In other words, a single electron fired at a time behaves as if it was interfering with itself in order to yield the interference on the screen.

Figure 19. Young's double-split experiment

Source: Stephen W. Hawking and Leonard Mlodinov, *The Grand Design*
(London: Bantam Press, 2010), 84.
This figure is reproduced courtesy of Professor Stephen W. Hawking, University of Oxford.

In a later variation of the double slit-experiment, the interference takes on the form of two intersecting cylindrical wavefronts. This happens, for example, when the two slits are illuminated by a single laser beam. The wave-like property of the subatomic particle directly speaks to imagination, for its interference pattern resembles the undulation of two expanding water waves that cross each other. Yet, there are also experiments in which the wave-interference is suppressed, so that only energy packets arrive at the screen. This is the case when particle detectors are positioned at the slits. With the use of particle detectors the experimenters are able to make out whether the particle (a photon or an electron) went only through one of the slits, or through both, as a wave would be expected to do. Depending on the apparatus used in experiments subatomic particles yield "packet"-results or wave-results, but never both at once.

This brings us to the heart of the Copenhagen interpretation; it tends to deny a proper reality to the quantum world independently of its being measured. Through our measurements we are, in a sense, the creators of the quantum reality that otherwise would remain nebulous: "The fuzzy and nebulous world of the atom only sharpens into concrete reality, when an observation is made. In the absence of an observation, the atom is a ghost. It only materializes when you look for it."[11] This is the common view of Niels Bohr and Werner Heisenberg (1901–1976), the two protagonists of the Copenhagen school. For the rest, they place their own accents. Bohr launched the complementarity principle according to which an experiment can demonstrate the particle-like properties of matter, or its wave-like properties, but not both at the same time, whereas Heisenberg made headlines with his uncertainty principle. He formulated this principle for the first time in 1926: if one measures a particle's position, then its velocity becomes uncertain; if one measures its velocity, its position becomes unpredictable. For Heisenberg, this uncertainty does not flow from some inaccuracy in our measure instruments; "it is truly intrinsic to nature."[12]

Indeed, in order to determine the behavior of an orbiting electron about its nucleus, one will have "to measure its present position and velocity accurately."[13] Now, the obvious way to do this is to shower the particle with an amount of energy, which can be done experimentally with light photons. When some of the light waves are scattered, this indicates that the electron got hit, and its position is detected. One, however, also observes that, owing to the added amount of energy, the particle abruptly changed its velocity. So, no prediction about its future behavior can be made in spite of, or precisely due to, the measurement. Moreover, if one would try to measure the particle's position more accurately, one would need to bombard it with light of shorter wavelength (the particle is detected then between the shorter wave crests). But this procedure still aggravates the situation:

> [for] the more accurately one measures the position, the shorter the wavelength of the light that one needs and hence the higher the energy of a single quantum [that is going to hit the electron]. So the velocity of the particle will be disturbed

11   Paul Davis, *God and the New Physics* (London: Penguin Books, 1983), 103.

12   Ibid., 102.

13   Stephen Hawking, *A Brief History of Time*, 58.

by a larger amount. In other words, the more accurately you try to measure the position, the less accurately you can measure its speed, and vice versa.[14]

Measurements usher in the randomness that is so typical of the quantum world: Observed particles just pop up, reveal part of their being, and disappear again by changing their velocity and wavelength.

When looking back at his earlier achievement, Bohr must have been shocked to have to admit that, from the standpoint of the new quantum theory, it becomes impossible to simultaneously determine an electron's location and its rotational velocity (its momentum). Uncertainty replaces precise prediction. That is the reason why, in the new quantum mechanical approach, instead of using the term "orbits," one prefers to speak of "orbitals," or "clouds" of possible locations of the electrons.

Probability Calculus: The Schrödinger Equation

The only real information one gets from the sub-atomic ghost world is through experiments. Yet, it is impossible to accurately predict what one is going to observe. This creates the need for engaging in probability calculus.

> In general, quantum mechanics does not predict a single definite result for an observation. Instead, it predicts a number of different possible outcomes and tells us how likely each of these is. That is to say, if one made the same measurement on a large number of similar systems, each of which started off in the same way, one would find that the result of the measurement would be A in a certain number of cases, B in a different number, and so on. One could predict the approximate number of times that the result would be A or B, but one could not predict the specific result of an individual measurement.[15]

This probability calculus is practiced in the various schools of quantum mechanics, including the Copenhagen school. But in 1926 the Austrian physicist Erwin Schrödinger developed a model of particle wave function that, in a rather classical form, allows us to predict how an accumulation of probabilities is going to evolve in time.

For Schrödinger, "the particles do not each have individual descriptions, but must be regarded as complicated superpositions of alternative arrangements of all of them together."[16] His point of departure is that

---

14   Ibid., 59.

15   Ibid.,60.

16   Roger Penrose, *The Emperor's New Mind*, 227.

there is a large number of alternatives available to every single position that a particle might have. So, what ought to be done is to mathematically combine all these available alternatives. This procedure takes on the form of complex-number weighting, in which both real and imaginary numbers are involved (Imaginary numbers give negative numbers when multiplied by themselves: e.g., $\sqrt{-2}$). Now, it is this collection of complex weighting that gives us the quantum state of the particle, that is: its wave function, indicated by the Greek letter ψ (pronounced "psi"). "For each position x, this wave function has a specific value, denoted $\psi(x)$, which is the amplitude [probability] for the particle to be at x."[17]

It is important that we realize the difference between probability amplitude (as a quantum effect) and the common notion of probability.

> When you first encounter the probabilistic aspect of quantum mechanics, a natural reaction is to think that it is no more exotic than the probabilities that arise in coin tosses or roulette wheels. But when you learn about quantum interference, you realize that probability enters quantum mechanics in a far more fundamental way. In everyday examples, various outcomes—heads versus tails, red versus black, one lottery number versus another—are assigned probabilities with the understanding that one or another result will definitely happen and that each result is the end product of an independent, definite history [...] But in quantum mechanics things are different. The alternate paths an electron can follow from the two slits to the detector screen are not separate, isolated histories. The possible histories commingle to produce the observed outcome. Some paths reinforce each other, while others cancel each other out. Such quantum interference between the various possible histories is responsible for the pattern of light and dark bands on the detector screen. Thus, the telltale difference between the quantum and the classical notion of probability is that the former is subject to interference and the latter is not.[18]

The probability amplitude is not really like a probability after all, if one takes the process of cancelations and reinforcements into account, but rather like a "complex square root" of a probability.

The Schrödinger equation, now, tells us how the wave-function actually evolves in time: the wave function ψ will not remain in its position state, but will disperse rapidly. Yet, "the *way* in which it disperses is completely fixed by this equation. There is nothing indeterminate or probabilistic about

---

17   Ibid., 243.

18   Brian Greene, *The Fabric of the Cosmos. Space, Time and the Texture of Reality* (London/ New York: Penguin Books, 2004), 208–9.

its behavior."[19] The Schrödinger equation gives us "a *completely determin-istic* evolution of the wave-function once the wave-function is specified at any one time!"[20] This leads us to the controversy that up till now continues to divide the scientific community. Whereas Niels Bohr thought that—inde-pendently of measurements— the quantum world had no proper reality of its own, Erwin Schrödinger opined that the wave function of a system does have physical reality, be it of a special kind. Or as Roger Penrose, a staunch proponent of the "physical reality" view, puts it:

> Many physicists, taking their lead from the central figure of Niels Bohr, would say that there is *no* objective picture at all. Nothing is actually "out there," at the quantum level. Somehow, reality emerges only in relation to the results of "measurements." Quantum theory, according to this view, provides merely a calcu-lational procedure, and does not attempt to describe the world as it actually "is." [Yet, Schrödinger] attributes *objective physical reality* to the quantum description— the *quantum state.*[21]

It is this "reality-geared" focus that motivated Schrödinger to underline the deterministic character of a particle's wave function in a probabilistic context.

The "collapse of the wave function" is a case in point of what is at stake in the controversy. In the wave function collapse, the initial superposition of several different states appears to be reduced to a single one of those states after interaction with an observer. Indeed, whenever one makes a "measurement" one magnifies the quantum effects to the classical level of observation: one jumps from the level of probability amplitudes, in which the mysterious interference happens, to the magnified level of observa-tion, where the interference stops and the wave function collapses. Penrose illustrates this with the help of a thought experiment that slightly modifies the double-slit scenario by placing a particle detector at one of the slits: "Since when it is observed," he writes, "a photon—or any other particle— always appears as a single whole and not as some fraction of a whole, it must be the case that our detector detects either a whole photon or no photon at all. However, when the detector is present at one of the slits, so that an observer can *tell* which slit the photon went through, the wavy interference pattern at the screen disappears. In order for the interference

19   Roger Penrose, *The Emperor's New Mind,* 251.
20   Ibid., 250.
21   Ibid., 226.

to take place, there must apparently be a 'lack of knowledge' as to which slit the particle 'actually' went through."[22] This again shows that one ought to regard the particle/wave as being spread out spatially, rather than being always concentrated at single points. Only in this spatially "spread-out" view does it make sense to speak of the evolution-in-time of a wave function; only then are we able to grasp that a "particle" can pass through the two slits at once, or that pairs of particles may simultaneously interact the one with the other even when they are light years apart.

The EPR Paradox

During his whole life time Albert Einstein refused to accept the premises of the Copenhagen school. For him, God "does not play dice"; there must be a more stringent order of cause and effect in the universe than Bohr and Heisenberg were willing to admit. In order to refute Bohr's interpretation that the specific behavior of the quantum world did not exist before it was observed, he set up with his associates Boris Podolsky and Nathan Rosen the paradox that is known after their initials: the EPR-paradox; it was published in 1935 with the title "Can Quantum Mechanical Description of Physical Reality Be Considered Complete?"

The essence of the EPR experiment is to examine the decay of an unstable elementary particle (an electrically neutral one, for example) into two photons. When the decay occurs, the photons will move off in opposite directions with equal momenta [i.e., with the same velocity] in order that the total momentum is conserved. These photons will also both possess an intrinsic *spin*,[23] and if one spins clockwise then the other must spin anticlockwise in order to conserve the total spin of the system during the decay process. By considering the complementary attribute to this spin, EPR were able to produce what they regarded as a paradox that demonstrated that quantum theory could not be a complete description of what was occurring. When the particle decays we cannot predict which residual photon will be found spinning in the clockwise direction. There is an equal probability for either the clockwise or the anticlockwise direction. But even if we let them fly apart to opposite sides of the universe, as soon as we measure the spin of one of them we should know, without any shadow of doubt, that the spin of the other will be found to be equal and opposite to the one we have just measured. This we will know *without* having measured it. Thus, EPR claims that the unmeasured spin of the other photon must be regarded as *real* according to their definition of physical

22   Ibid., 236.
23   For a detailed analysis of the spin notion, see the next section.

reality, because it is *predictable*. It must, therefore, be related to some element of an observer-independent reality.[24]

EPR made a point. It is possible to determine (without measuring it) the spin value of photon B moving with the same velocity in the opposite direction by only measuring the spin value of photon A; for on the basis of the conservation of the total spin of the system one can tell that when photon A has a clockwise spin, photon B must have an anticlockwise spin, and vice versa. So, there is something objective in the quantum world, which is observer-independent. Yet, EPR also points to an inherent problem: how can photon B know that photon A took on a clockwise spin during measurement and accordingly adopt a counterclockwise spin? How can photon B seemingly simultaneously react to the new event, even when it finds itself at the other side of the universe? Such an instantaneous interaction clashes with the relativity principle that nothing can travel faster than the speed of light, 300,000 km/sec.

In the meantime, however, this instantaneous communication, technically called "quantum nonlocality," has been confirmed:

> It looks as if spatial distances do not exist. Einstein and his associates claimed that such a nonsensical result demonstrates the fact that quantum theory is not a complete theory. Yet, what in Einstein's time could only be a thought experiment, nowadays, with our present technology, has been many times performed. And it has turned out that the "crazy" prediction of quantum mechanics is correct. "Quantum nonlocality," as it is now called, is an empirical fact, and physicists have to live with it [...] It is only our intuition that cannot easily accept that two elementary particles can interact with each other with no mediation of space and time distances separating them.[25]

When commenting on the EPR paradox, Penrose comes to the same conclusion. For him too, it is evident that

> the measurement of an entangled system [i.e., of photon A and its twin photon B] sits extremely uncomfortable with the requirements of relativity, since a measurement of one part of an entangled pair would seem to have to affect the other

---

24   John Barrow, *The World Within the World*, 146.

25   Michael Heller, "Where Physics Meets Metaphysics," in *On Space and Time*, ed. Shahn Majid (Cambridge: Cambridge University Press, 2008), 260.

simultaneously, which is not a notion that we ought to countenance if we are to remain true to the relativistic principles.[26]

What makes quantum physics incompatible with Einstein's theory of special relativity is precisely the phenomenon of entanglement. Quantum mechanics conceives of two photons moving outwards in space as "photon *pairs,* acting as a single unit. Neither photon individually has an objective state: the quantum state applies only to the two together."[27] It is this entanglement that enables the twin pair to instantaneously interact when one of them is "forced" by a measurement to take on a certain spin value. All this happens as if the photon pair, even when they are light years apart, was not in the least encumbered by the classical notion of spacetime. Entanglement of twin photons is an element that Einstein could not have foreseen.

Pauli Exclusion Principle

Up till now we have mentioned only the elementary particles electron, proton, neutron, and photon. Yet, the more new particles were discovered, the stronger the need to classify them. The earliest classification was that of leptons (*leptos* in Greek means "light," not heavy) and baryons (*barus* in Greek means heavy). The electron is a lepton, whereas proton and neutron fall under the category of the baryons. Leptons cannot further be divided in subcomponents, but baryons can. As we will see below, proton and neutron (which are baryons) are made up of quarks. That is the reason why recently one has begun to refer only to two types of basic constituents of mater: leptons and quarks. Ever since the discovery of quarks each composite particle made up of quarks and held together by the strong force is called a hadron (*hadros* in Greek means bulky). Hadrons are categorized into two families: baryons (see above) made up of three quarks, and mesons (from the Greek *mesos:* medium) made up of one quark and one antiquark. The best-known baryons are protons and neutrons; the best-known mesons are the pions and kaons, which were discovered in cosmic ray experiments in the late 1940s and early 1950s.

The above classifications mainly deal with matter particles, with the neglect of energy particles. Leptons, quarks, baryons, and mesons are matter particles, also termed fermions, named after Enrico Fermi. But

---

26   Roger Penrose, *The Road to Reality: A Complete Guide to the Laws of the Universe* (London: Jonathan Cape, 2004), 593.

27   Roger Penrose, *The Emperor's New Mind,* 287.

besides them, there also exist energy particles or particles that are carriers of forces (the electromagnetic force, the strong and the weak nuclear force, and gravity). The force-carrying particles are termed bosons, derived from the name of Satyendra Nath Bose. All the fundamental particles in nature fall into these two categories: they are either fermions or bosons. The table below lists the differences.

Article:	Spin:	Occupancy of a State:	Examples:
Fermions	Half-integral	One	electrons, protons, neutrons, quarks, neutrinos
Bosons	Integral	Many	photons, He atoms, gluons

Figure 20. Bosons and fermions

A look at the table shows that the basic difference between fermions and bosons consists in their specific spin. Spin is a measure of a particle's rotation about its axis; it is an *intrinsic* property of the particle (i.e., not arising from some orbital motion about a center). Fermions have half-integral spin, whereas bosons have integral spin, that is: fermions, whatever they are doing, perform time and again half a rotation about their axis, whereas bosons perform time and again a whole rotation about their axis.

There is still another striking feature. A particle's spin either takes on a "spin up" value (rotating upward anti-clockwise) or a "spin down" value (rotating downward clockwise). We have seen this phenomenon at work in the EPR paradox: there it became evident that in the case of two photons moving off in opposite directions one had to have a clockwise spin ("spin down") and the other an anticlockwise spin ("spin up"). A differentiation like this not only occurs with bosons (a photon is a boson); it is also typical of the fermions, for which the divide between "spin down" and "spin up" is, in a sense, required to make the Pauli Exclusion Principle work.

The Pauli Exclusion Principle was formulated by the Austrian physicist Wolfgang Pauli in 1925. It states that no two *identical* fermions (with half-integral spin) may occupy the *same* quantum state simultaneously. The insight into this principle flows from statistic data. Bosons have the tendency to clump into the same quantum state, thus fortifying the strength of their common effect, as this is the case with laser beams, whereas fermions behave as if they were forbidden from sharing identical quantum states, and thus, from blending.

Pauli used the Exclusion Principle to explain the arrangement of electrons in an atom. In the quantum mechanical approach to atoms, the space surrounding the dense nucleus is thought of as consisting of orbitals, or regions in which it is most probable to find electrons. According to the Pauli Exclusion Principle, no two electrons with the *same* quantum state can exist within the *same* orbital. This implies that *each* electron in an orbital has a *unique* quantum state, described by a *unique* set of quantum numbers. The principal quantum number indicates the energy level of the orbital, a second number represents the shape of the orbital; a third number indicates the orbital angular momentum, and a fourth indicates the particle's spin direction ("up" or "down").

Two electrons with the same energy level and orbital angular momentum can, to be sure, exist in the same orbital, but it will be necessary that they have opposite spins: one with "spin up" the other with "spin down." The opposite spin differentiates them, even when for the rest they share the same properties. In this way, the Pauli Exclusion Principle keeps the distinct electron waves apart, not allowing them to merge. If such a merger were to happen, the atom in question would degenerate and fall apart. The Exclusion Principle applies to all fermions, and thus also to quarks: "If the world had been created without the exclusion principle, quarks would not form separate, well-defined protons and neutrons. Nor would these, together with electrons, form separate well-defined atoms. They would all collapse to form a roughly uniform, dense 'soup.'"[28] In other words, without the Exclusion Principle no distinct forms of matter would ever have come into existence; the universe would still be an undifferentiated "soup" as it most probably was "in the beginning," right after the Big Bang.

We cannot conclude this section without expounding on the nature and properties of the force-carrying particles: the bosons with spin 0, spin 1, and spin 2. They all regulate the interactions between matter particles, and do not obey the Pauli Exclusion Principle. So, they can pile up and increase their strength. "Force-carrying particles can be grouped into four categories according to the strength of the force that they carry and the particles with which they interact."[29]

(i) *The gravitational force,* which is located in the gravitational field, where it acts between matter particles. This force is rather weak, "but it can act over large distances, and it is always attractive [...] it is pictured as

---

28   Stephen Hawking, *A Brief History of Time,* 72.

29   Ibid., 73-74.

being carried by a particle of spin 2, called the *graviton;*" the existence of the graviton, however, is not yet experimentally verified. (ii) *The electromagnetic force,* which interacts with electrically charged particles. It "is pictured as being carried by massless particles of spin-1, called *photons."* (iii) *The weak nuclear force,* which is responsible for radioactivity; it is carried "by three spin-1 particles, known collectively as *massive vector bosons* [...] These bosons are called W^+ (pronounced W plus), W^- (pronounced W minus), and Z^0 (pronounced Z naught)." Because these particles are so massive their action radius is extremely short. (iv) *The strong nuclear force,* finally, which holds the quarks together in the proton and neutron and holds the proton and the neutron together in the nucleus, is of a comparably short range: "It is believed that this force is carried by another spin-1 particle, called the *gluon,* which interacts only with itself and with the quarks."[30] This is why the more one would try to separate the quarks, in a neutron e.g., the more forcefully the gluon would react in keeping the quarks together.

electromagnetic force	weak interaction	chromodynamics strong nuclear interaction	gravity
photo	bosons W+, Zo	gluons	graviton?
mass = 0	mass = 80/90	mass = 0	mass = 0

Figure 21. The four forces and their mediator bosons

The Discovery of Antiparticles

The prediction that each particle would be accompanied by an anti-particle was made by the British physicist Paul Dirac (1902–1984). In 1928 he elaborated an extension of the Schrödinger equation, a new wave equation that is consistent with special relativity.

(i) In retrospect, it is easy to understand that the existence of antiparticles logically flows from Einstein's formula $E = mc^2$. Nowadays we know that if a particle is hit by its antiparticle, they annihilate each other so that energy is released in the form of photons. Conversely, "if sufficient energy is introduced into a system, localized in a suitably small region, then there arises the strong possibility that this energy might serve to create some

---

30   Ibid., 74–77.

particle together with its antiparticle."[31] Yet, the $E= mc^2$ formula alone cannot give us a clue as to why for every kind of particle there is a corresponding antiparticle with opposite electric charge. It is here that quantum physics comes into play. According to this theory, a particle-antiparticle pair can only be created on condition that there is no violation of charge conservation: when a particle has a specific quantity of *negative* electric charge, e.g., its antiparticle must have the same quantity of *positive* electric charge.

(ii) I have given this retrospective view in order to show how difficult it is to harmonize quantum mechanics with Einstein's special relativity. Dirac had to wrestle a great deal with such intrinsic difficulties. In his attempt at bringing the Schrödinger equation in line with special relativity, he met with an unsettling problem. The new equation made it clear that for each quantum state with positive energy E, there was a corresponding state with energy "minus E." In Einstein's approach, there is no place for negative energy, that is: for an energy state below zero. Yet, "in quantum mechanics, one has to consider that the various possible things that 'might' happen, in a physical situation, can all contribute to the quantum state [...] So, even an 'unphysical negative energy' has to be considered as a 'physical possibility.'"[32]

Facing up to this difficulty, Dirac set out to imagine a huge pool of negative-energy electrons, and then wondered what would happen if a positive-energy electron[33] would tumble into this pool. Such a positive-energy electron would without doubt be absorbed into the "sea" of negative-energy electrons, and the more it would immerse in it, the more it would shed its positive energy by emitting photons. Yet, if an unlimited number of electrons were to do the same, a catastrophic instability would arise as a result of the limitless release of energy. But apparently such instability has never occurred. To prevent this situation from happening, Dirac hypothesized that a "sea" of negative-energy electrons fills the universe in such a way that all of the lower-energy states were occupied. So, in line with the Pauli Exclusion Principle, no other electron could fall into the "sea." What remained possible, however, is that one of the negative-energy particles could be lifted out of the "sea," thus leaving behind a *hole* that would act like a positive-energy electron with a reversed charge.

31   Roger Penrose, *The Road to Reality*, 610–11.

32   Ibid., 615.

33   Note that an electron with negative electric charge has positive energy.

If then, Dirac continued, a negatively-charged electron would fall into the "hole," the effect of it would be flabbergasting, "it would result in the 'hole' and the electron annihilating one another, in the manner that we now understand as a particle and its antiparticle undergoing mutual anni-hilation,"[34] leading to a bursting forth of energy in the form of photons. Dirac's "hole" is indeed the electron's antiparticle.

Initially Dirac thought that the "hole" could be a proton. When he was conducting his research, it was commonly held that protons were the only particles thought to have a positive electric charge. This guess turned out to be erroneous, since the mass of the proton is about 1836 times larger than that of the electron, and the mass of a particle and an antiparticle, respectively, had to be equal. That's why Dirac in 1931 termed the predicted particle an anti-electron, an unknown particle that we now call a positron, the existence of which was, the very next year, effectively discovered by Carl Anderson.

Quantum Field Theories

In the previous pages our primary focus has been on Quantum Mechanics. This branch of quantum physics refers to a system in which the number of particles is fixed and the fields, such as the electromagnetic field, are continuous classical entities. Quantum Field Theories (QFTs), however, move beyond this fixed view in that they analyze how particles are incessantly being created and annihilated

Dirac can be regarded as a precursor of QFTs because he already reck-oned with the impact of electric fields on the formation of quanta. Yet, it is only a reformulation of his equation that brings us into the proper domain of QFTs. Dirac still saw his equation as depicting the evolution-in-time of one single electron, whereas, in fact, it should be read as the evolution-in-time of a whole quantum field in which operators cause the creation and annihilation of particles. Operators are mathematical entities that are defined by how they act on something; they bring about transformations in the states of the field, for example, excite it in a particular mode at a particular point x. Depending on these excitations particles can be created or annihilated.

Quantum Field Theories inaugurate a new era of research in particle physics, as well as establishing the basis for further rapprochement between quantum physics and special relativity. This rapprochement can be seen, for example, in the central position the notions "symmetry" and

---

34   Roger Penrose, *The Road to Reality*, 625.

"invariance" occupy in the theory formation of QFTs: "Symmetries are perhaps the most important objects in all of mathematics."[35] In geometry, an object is symmetric with respect to a given operation, if this operation, when applied to the object (a geometrical shape), does not appear to change it. There are various "symmetry operations" that can be performed, such as translation (which "slides" an object from one area to another by a vector), or rotation around a fixed central point x. In physics, however, symmetry has been generalized to mean invariance—that is, lack of change—under any kind of transformation.

In QFTs the following symmetry operations are practiced: (i) C, which stands for *charge conjugation*: "This operation replaces every particle by its antiparticle [...] A physical interaction that is invariant under the replacement of particles by their antiparticles (and vice versa) is called C-invariant. (ii) P, which stands for *parity*: This operation replaces a particle by its mirror image (reflection in a mirror). A physical interaction that is invariant under the replacement of a particle by its mirror image is called P-invariant. (ii) T, which stands for *time reversal*. "An interaction is invariant under T if it is unaltered if we view it from the perspective of a time direction that is the reverse of normal," [36] i.e., when time runs back. In the three cases, one has to do with heuristic tools allowing the researchers to investigate whether, in a particular domain, a certain basic property of the system is, in actual fact, invariant. Moreover, the study of symmetry operations may reveal that under certain circumstances a symmetry break must have taken place (see below: Electroweak Theory).

The original title Einstein planned for his special theory of relativity was "invariant theory." In this theory, the invariant par excellence is the speed of light (300,000 km/sec). The speed of light is always the same, regardless of the time of measurement, regardless of the speed of the light source, and regardless of the speed with which the platform moves from which the speed of light is measured. Einstein's special theory of relativity is based on the implementation of symmetry operations. The type of symmetry he sought to establish is referred to as "global symmetry," that is, a symmetry in which the basic property (the speed of light) is invariant under spatio-temporal transformations at large.

Quantum field theories, on the contrary, are "local symmetries." Whereas global symmetries "were invariances that existed when you changed things

35   Shahn Majid, "Quantum Spacetime and Physical Reality," in *On Space and Time*, ed. ID (Cambridge: Cambridge University Press, 2008), 80.

36   Roger Penrose, *The Road to Reality*, 638.

in the same way everywhere, local gauge symmetry requires invariance after you have done different things in different places at different times."[37] In short, local gauge symmetries are predicated on the idea that symmetry transformations are performed locally in a particular domain of space-time. This allows for diversification, but also for the imposition of strict conditions. In Maxwell's equations, e.g., the local gauge symmetry "forbids the existence of a photon possessing a mass, and it dictates the precise way in which electrically charged particles interact with light."[38]

So, too, in Quantum Field Theories, "the *local* gauge symmetry dictates what mediating forces must exist between the particles involved."[39] These mediating forces are all bosons, particles with integer spin that are, with the exception of W+ and W-, themselves their own antiparticles. In Quantum Electrodynamics electrically charged particles interact with each other through an exchange of photons. For the weak nuclear force the mediating bosons are W^+, W^-, and Z^0, whereas in the strong nuclear force gluons mediate between the quarks. For the force of gravity, finally, the graviton is postulated as the mediating boson. Similarly special mathematical symmetry groups are operational in the various Field Theories: the symmetry group U (1) in Quantum Electrodynamics; the symmetry group SU (2) in the Weak Nuclear Force, and the symmetry group SU (3) in the Strong Nuclear Force.

Quantum Electrodynamics

Quantum Electrodynamics (QED) was the first success story of Quantum Field Theory; it is an Abelian[40] gauge theory that operates with the symmetry group U(1). QED gives a highly satisfactory quantum description of the behavior of electrically charged particles in an electromagnetic field, including the creation and mutual annihilation of particle-antiparticle pairs. It is based on the idea that electrically charged particles interact with each other in the electromagnetic field by emitting or absorbing photons, the massless particles of spin 1 that transmit the electromagnetic force.

A case in point of this interaction is "the electromagnetic attraction between negatively charged electrons and positively charged protons in the nucleus, which causes the electron to orbit the nucleus of the atom. The electromagnetic attraction is pictured as being caused by the exchange

---

37   John Barrow, *The World Within the World*, 181.

38   Ibid.,181.

39   Ibid.,182.

40   Abelians are symmetry groups with the commutative property *ab=ba*.

of a large number of massless particles of spin 1, called photons."[41] These exchanged photons are virtual particles that, because of their confinement in atoms, can hardly be detected. They can, however, become real photons, when released from the atom, which happens when an electron jumps from a higher allowed orbital to a lower one. In that case it can freely travel though space.

But more spectacular is the way in which electron-positron pairs annihilate one another by emitting two photons. In the diagram below, an electron (e⁻) and a positron (e⁺) meet and annihilate each other, so as to give rise to two photons (γ).

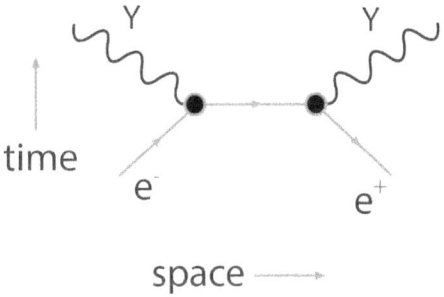

Figure 22. Electron-positron annihilation

Source: Wikipedia (http://en.wikipedia.org/wiki/Electron%E2%80%93positron_annihilation),
accessed September 23, 2014.

In 1948 the American physicist Richard Feynman brought the various interactions between electrically charged particles and photons into picture with his Feynman diagrams. These diagrams depict various constellations in which electrons and positrons travelling independently through space and time may collide so as to annihilate and emit two photons; conversely they may illustrate how a photon pair gives rise to the creation of an electron-positron pair. The aim and purpose of these various diagrams is to offer a practical tool for calculating the probability with which, in a particular constellation, the "photon exchange" takes place, and with which intensity it may happen. In Feynman's approach "a particle does not just have a single history, as it would in a classical theory. Instead, it is supposed to

---

41   Stephen Hawking, *A Brief History of Time*, 75.

follow every possible path in space-time,"[42] and to interact in every way available. The probability of each final state is then obtained by summing over all the possible paths.

In the beginning QED was still plagued with infinities, which indicated that no absolutely precise result was obtained. The difficulty was overcome in the late 1940s with the development of a procedure called renormalization. "It consisted of the rather arbitrary subtraction of certain infinite quantities to leave finite remainders. In the case of electrodynamics it was necessary to make two such infinite subtractions, one for the mass and one for the charge of the electron."[43] Ever since, renormalizability has become a prerequisite for a theory's recognition in the scientific community: "QED is a renormalizable theory. [...] It is a common standpoint, among particle physicists, to take renormalizability as a selection principle for proposed theories. Accordingly, any non-renormalizable theory would be automatically rejected as inappropriate to Nature."[44]

Electroweak Theory

Quantum Electrodynamics lies at the basis of the Standard Model, which comprises also the study of the weak and the strong nuclear force. The major success of the Standard Model was the unification of the electromagnetic force and the weak nuclear force. Such unification, dubbed Electroweak Theory, is not simple, for the two forces have on prima facie consideration very little in common. First of all, the weak nuclear force acts across distances smaller than the atomic nucleus, while the electromagnetic force can extend for great distances in space-time (think of the light of stars reaching us from other galaxies). And second, the weak force is some 1,000,000 times weaker than the electromagnetic force.

The study of the weak force is aimed at developing a consistent gauge theory that could withstand comparison with Quantum Electrodynamics. Weak interactions are responsible for the decay of fundamental particles into other (lighter) particles: one observes the particle vanishing and being replaced by two or more different particles. In beta minus decay, for instance., a neutron decays into a proton, an electron, and an antineutrino.

During the 1960s Sheldon Glashow, Abdus Salam, and Steven Weinberg independently of one another discovered that a gauge theory of the weak

---

42   Ibid, 141.

43   Stephen Hawking, "Is the End in Sight of Theoretical Physics?," in *Stephen Hawking's Universe*, ed. John Boslough (New York: Avon Books, 1980), 125–26.

44   Roger Penrose, *The Road to Reality*, 678.

force could be constructed, if the electromagnetic force was included in the theory formation. This inclusion would provide the researchers with a platform from where to build an appropriate gauge theory. A major obstacle, however, in developing such a theory was the fact that the W^+, W^- and Z° bosons that regulate the weak interaction are extremely massive, while photons, the bosons of the electromagnetic force, are massless. The heavy W^+, W^- and Z° particles are accurately described by the mathematical symmetry group SU(2), whereas according to the mathematical symmetry group U(1) that sets the tone in Quantum Electrodynamics the photon must be massless. So, a new special symmetry device will be needed to unite the two previous ones, namely the symmetry group SU(2) x U(1). It is this symmetry group that supplies the underlying pattern for the way in which the bosons of the unified electroweak force are fully interrelated: "All of W^+, W^-, Z^0, and γ can, in a certain sense, be continuously 'rotated into one another,'"[45] so that all of them are, basically, on the same footing.

The above unification only works under special conditions. The key issue is, indeed, that the four mediator particles not only have different electric charges—the photon γ, and Z^0 have no electric charge, whereas W^+ and W^- have positive and negative charge—but also different masses: the photon particle γ is massless, whereas W^+, W^-, and Z^0 are extremely massive (mass 80/90). So how can the theory claim that the electromagnetic and the weak force behave as if they were basically the same? Facing up to this difficulty, Glashow, Salam, and Weinberg assumed that the reconstructed "electroweak" force is only workable under extreme temperatures (or, high energy scales), as these existed in the early universe. So, the asymmetry we observe today would be "the result of a *spontaneous* symmetry breaking that is taken to have occurred in the early stages of the universe. Before that period, conditions were very different from those holding today, and the standard electroweak theory asserts that in the extremely high temperatures in the early universe the U(2) symmetry [that is: the SU(2) x U(1) symmetry] held *exactly*," so that W^+, W^-, Z^0, and γ could be rotated into one another.

But, as the idea goes, when the temperature in the universe cooled (to below 10^{16} Ke,[46] at about 10^{-12} seconds after the Big Bang), the particles W^+, W^-, Z^0, and γ were "frozen out" by this process of spontaneous symmetry breaking. [...] Just three of them acquire mass, and are referred to as the Ws and the Z°; the other one

---

45   Ibid., 641.
46   One degree Kelvin (Ke) equals minus 273.15 degrees Celsius.

remains massless and is called the photon. In the initial "pure" unbroken version of the theory, when there was complete U(2) symmetry, the Ws, Z^0, and γ would all have to be effectively massless.

In order to explain the mechanism of symmetry-breaking the three scientists had recourse to an additional interaction with another particle/field, known as *Higgs* (particle): "The Higgs (field) is regarded as being responsible for assigning mass to all these particles (including the Higgs particle itself) and also to the quarks that compose other particles in the universe."[47]

The massive character of the weak gauge bosons W^+, W^-, and Z^0 can be derived from the fact that the weak force is short range, which reveals that it is carried by massive particles. In 1971 Gerard 't Hooft proved that the unified electroweak theory proposed by Glashow, Salam, and Weinberg was renormalizable, and the theory gained full respectability. In 1983 W^+, W^-, and Z^0 were discovered at CERN (European Centre for Nuclear Research, near Geneva, Switzerland), "with the correct predicted masses and other properties."[48] Glashow, Salam, and Weinberg received the Nobel Prize in physics in 1979 even before the predicted particles were detected in the Geneva particle accelerator. The discovery of the Higgs field/particle figured as a top priority on the agenda of the Large Hadron Collider that was successfully put to use by CERN in March 2010. In July 2012 scientists at CERN discovered traces of the Higgs field/particle. In 2013 François Englert, together with Peter Higgs, received the Nobel Prize in physics for the discovery of the Higgs mechanism.

Quantum Chromodynamics

The successful reconstruction of the electroweak force was one of the major achievements of the Standard Theory, soon to be followed in the 1970s by the elaboration of Quantum Chromodynamics (QCD), the gauge theory that describes the working of the strong nuclear force. According to this theory, the strong nuclear force is transmitted by bosons dubbed gluons. Like photons, gluons are massless, have a spin of 1, and travel at the speed of light. But they differ from photons in one important respect: they carry what is called "color" charge, a property analogous to electric charge but which has almost no manifestation at distances above the size of the atomic nucleus. QCD is modeled after Quantum Electrodynamics.

---

47   Ibid., 643.

48   Stephen Hawking, *A Brief History of Time*, 77.

However, whereas in Quantum Electrodynamics the exchange of photons took place between *two particles* with opposite electric charge (namely electron and positron), QCD had to elaborate an exchange of gluons between *three colored quarks* (see detailed analysis below). This change in perspective relates to the fact that it was discovered that proton and neutron were not indivisible as was thought till then, but made up of three types (flavors) of quarks.

For a long time attempts had been made to explain the way in which protons and neutrons were held together in the atomic nucleus. In the 1930s it was suggested that the force that bound them together was produced by a pion (a meson with a mass intermediate between that of the electron and the proton). Yet, since the late 1940s a vast number of heavy particles (hadrons) —Λ^0, Σ^0, Ω^- etc. —had been discovered in cosmic rays and accelerators which all fell into certain families, called *multiplets*. This discovery led to the insight that the heavy particles that feel the strong nuclear force[49] were made up of "families" of smaller particles for which the American physicist Murray Gell-Mann coined the name "quarks." He took the designation "quark" from James Joyce's novel *Finnegans Wake*, in which the whimsical saying appears: "Three quarks for Muster Mark." In his study of the strong nuclear force Gell-Man worked with the SU(3) symmetry group and came to the conclusion that protons and neutrons (which are hadrons) "were made up of three types of quarks referred to as three *flavors*, rather unimaginatively called: 'up,' 'down,' and 'strange.' A mysterious feature of quarks was that they have to possess fractional electric charge [...], the up, down, and strange quarks having respective charge values ⅔, -⅓, and -⅓."[50]

A proton contains two "up" quarks and one "down" quark (+⅔ +⅔ -⅓) giving a total electric charge of +1, whereas a neutron is made up of two "down" quarks and one "up" quark (-⅓ -⅓ +⅔), giving a total electric charge of zero. So, the fractional electric charge poses no problem at all, since the respective arrangements thereof result in a proton with a positive electric charge, and a neutron with a zero electric charge, as this is required by the atomic structure.

Yet, on close inspection, these electric charge arrangements create a problem. Quarks are matter particles (fermions), so they all have half-integer spin, and consequently must obey the Pauli Exclusion Principle

---

49   The same insight applies to mesons: see below.

50   Roger Penrose, *The Road to Reality*, 646.

Quark	Symbol	Spin	Charge	Baryon Number	Mass*
Up	U	½	+2/3	1/3	1.7-3.3 MeV
Down	D	½	-1/3	1/3	4.1-5.8 MeV
Charm	C	½	+2/3	1/3	1270 MeV
Strange	S	½	-1/3	1/3	101 MeV
Top	T	½	+2/3	1/3	172 GeV
Bottom	B	½	-1/3	1/3	4.19 GeV(MS) 4.67 GeV(1S)

Figure 23. Types of quarks

that forbids *identical* fermions from sharing the *same* quantum state. This principle, however, is not being respected: Indeed, in the configuration that forms the proton (+⅔ +⅔ -⅓) we have two "up" quarks with the same electric charge +⅔, whereas the configuration of the neutron (-⅓ -⅓ +⅔) includes two "down" quarks with the same electric charge -⅓. This boils down, in both cases, to a violation of the Pauli Exclusion Principle. This raises the thorny question as to whether quarks are really existing entities or just a matter of bookkeeping to ensure the correctness of the mathematical calculations proper to the symmetry group SU(3). As a matter of fact, the bookkeeping only works if one pretends that quarks are "bosons," to which the Pauli Exclusion Principle does not apply. Yet, such a trick is, after all, not really convincing.

In order to resolve this problem the researchers set out to postulate that, in line with the triplet preference of the symmetry group SU(3), "each flavor [i.e., type] of quark also comes in three (so called) 'colors.'"[51] Color stands for "color charge." The diversified quarks are taken to have "red," "green," or "blue" color charge; these are combined in such a way as to yield a colorless composite particle (red +green + blue = white). On the basis of this procedure, the problem raised by the Pauli Exclusion Principle is resolved: in the arrangement of the proton (+⅔ +⅔ -⅓), e.g., we have, to be sure, two "up" quarks with the same electric charge +⅔; yet, these two "up" quarks are split up now into "up red" and "up green." In this way, no two identical quarks share the same quantum state; in combination with a

---

51   Ibid., 648.

blue "down" quark, with electric charge -⅓, they form together a colorless proton.

The theorem of quark colors has still a further advantage: it allows for a deeper insight into the very nature of the gluons and their interaction with quarks. Researchers found out that gluons carry with themselves color charge, in sharp contrast to the photons, the exchange particles in Quantum Electrodynamics, which have no electric charge of their own. Gluons, the bosons of the strong force, can, in other words, radiate an array of gluons, whereas photons are not able to increase their number. This radiation generates a "sea" of virtual gluons with the effect that the quarks in this "sea" engage in a dynamic interaction both with each other's color charge and with the gluon's intensified color charge. It is in this context that one has to place the phenomenon of color change. When a gluon is transferred between two quarks, both quarks change their color: a red quark that emits a red–antigreen gluon becomes green, whereas a green quark that absorbs a red–antigreen gluon becomes red. Besides the quarks' color change, there is also the phenomenon of confinement. The more quarks try to escape their confinement within the proton or the neutron, the more the "sea" of virtual gluons grows in strength. At tiny distances from the center, quarks behave as if they were nearly free to move around. However, when one begins to draw the quarks apart in an attempt at knocking them out of a proton, the binding force grows stronger. This phenomenon is in complete contrast with the electric and gravitational force between particles whose effects become weaker with the square of the distance: "The strong force is more like an elastic band, where the strength of the force increases in proportion to the distance of stretch, and it drops to zero, when the distance becomes zero." [52] It is this increase in force that is held responsible for the fact that quarks cannot be individually pulled out of a proton or neutron.

At the end of the Twentieth Century, particle accelerators were already strong enough to test the results of particle physics. Colliding beam experiments resulted in the discovery of the charm quark at SLAC (Stanford Linear Accelerator Center) in 1968. All six flavors of quark have since been observed in particle accelerator experiments; the top quark, first observed at Fermilab (Fermi National Accelerator Laboratory, Chicago) in 1995, was the last to be discovered. The gluon was first discovered in 1979 at the Deutsches Elektronen-Synchrotron (DESY) in Hamburg, Germany.

---

52   Roger Penrose, *The Road to Reality,* 679.

The strong force is not only at work in protons and neutrons (hadrons); it also plays a decisive role in the constitution of mesons, particles with a mass intermediate between that of the electron and the proton. Mesons, too, are made up of quarks, but for them the quarks come in pairs, which makes them less stable (quarks and antiquarks may annihilate one another). A "pi" particle, for instance, is made up of one "down antiquark" and one "up quark" ($+\frac{1}{3}$ $+\frac{2}{3}$) resulting in a total electric charge of +1, whereas a "kaon," which was first discovered in 1947, contains one "up quark" and one "strange antiquark" ($+\frac{2}{3}$ $+\frac{1}{3}$) giving an electric charge of +1. The mesons also possess color charge. A meson may consist of a red quark and an anti-red quark, or a green quark and an antigreen quark, or a blue quark and an antiblue quark. The result of all these combinations is, in each case, a colorless meson.

Further Prospects: Grand Unified Theories

The successful study of Quantum Chromodynamics encouraged researchers to explore the possibility of uniting the electroweak and the strong nuclear force in what is termed Grand Unified Theories (GUTs). Experiments with large particle accelerators have shown "that at high energies the strong force becomes much weaker, and the quarks and the gluons behave almost like free particles."[53] In normal circumstances, quarks and gluons behave only that way when moving close to the center of the protons and neutrons, whereas this freedom drastically diminishes as the distance from the center increases. Yet, at high energies—which also means: at extremely high temperatures—this freedom of movement is enhanced and extended to regions where quarks and gluons find themselves at considerable distances from the center, which means that, on the whole, the strong force gets weaker. This overall weakening renders the unification possible with the electroweak force.

> The basic idea of GUTs is as follows: As was mentioned above, the strong nuclear force gets weaker at high energies. On the other hand, the electromagnetic and weak forces [...] get stronger at high energies. At some very high energy, called the grand unification energy, these three forces would all have the same strength and so could just be different aspects of a single force. The GUTs also predict that at this energy the different spin ½ particles, like quarks and electrons, would also all be essentially the same, thus achieving unification.[54]

---

53   Stephen Hawking, *A Brief History of Time*, 79.
54   Ibid., 79.

Quarks, the constituents of protons and protons that possess color charge, would simply behave as electrons without color charge!

In the above scheme, gravity is not involved. An encompassing theory that would include gravity—the so called "Theory of Everything" (TE) —is

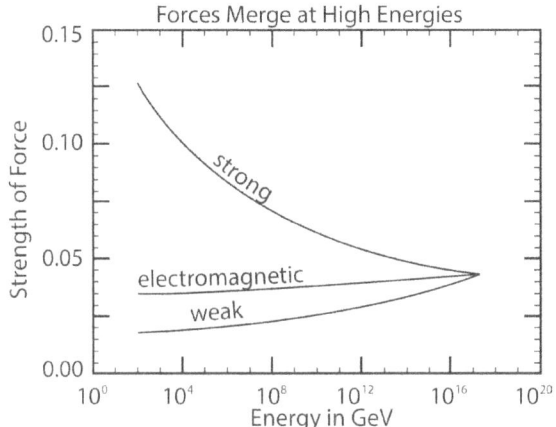

Figure 24. Forces merge at high energies

Source: The Particle Adventure (http://www.particleadventure.org/grand.html), accessed September 23, 2014.

not easy to obtain: "The main difficulty in finding a theory that unifies gravity with the other forces is that general relativity is a 'classical theory'; that is, it does not incorporate the uncertainty principle of quantum mechanics. On the other hand, the other partial theories depend on quantum mechanics in an essential way. A necessary first step, therefore, is to combine general relativity with the uncertainty principle."[55] There are various attempts at integrating the uncertainty principle into general relativity. One of them has come about as a result of the study of black holes.

The existence of black holes— "voids" in space—flows from the general theory of relativity, which predicts its own inability to determine what happens at a singularity. At a singularity, the theory of relativity breaks down. "A *singularity* is defined to occur when the path of a light ray, or that of a particle, comes to an end. If this happens, then, on reaching the end of its path the particle disappears from the universe because it runs

---

55   Ibid., 164–65.

out of space and time."[56] The "big bang," insofar as it is to be located before the formation of space-time, is a case in point of a singularity. So, too, is a black hole that emerges in the expanding universe: "If sufficient mass is attracted into a small enough region by the pull of gravity, then the gravitational field that it creates can become so strong that nothing can escape—not even light."[57] Any light or other signal that comes close to the black hole is dragged into it and cannot escape to the outside world.

Yet, in 1974 Stephen Hawking argued that the black hole must undergo the impact of entropy and that, therefore, it ought to have a temperature and emit radiation. He corroborated this view by showing that the gravitational field at the edge of the black hole (technically: at its "event horizon") must have quantum fluctuations with all that this involves: "One can think of these fluctuations as pairs of particles of light or gravity that appear together at some time, move apart and then come together again and annihilate each other."[58] The particles described are virtual particles. Now, it may happen, Hawking goes on, that particles with negative energy fall into the black hole and become real particles or antiparticles, while those with positive energy escape from the black hole into infinity, in the form of radiating energy. This baffling situation confronts us with a double unpredictability. If in quantum physics, in normal circumstances, one cannot measure at the same time the position and the velocity (momentum) of a particle, this uncertainty is doubled now by the fact that one is not able to tell what happens with the twin particles that fell into the black hole: "An observer at a distance from the black hole can measure only the outgoing particles, and he cannot correlate them with those that fell into the hole because he cannot observe them."[59] Particles of light or gravity can escape from the back hole, but what happens with their twin particles in the black hole remains uncertain. In this way, the uncertainty principle of quantum mechanics is given a place in the theory of relativity. Yet, this realization is only a modest step in the direction of a real incorporation of quantum mechanics into the theory of relativity. Hawking acknowledges that "we do not yet have a proper quantum theory of gravity, let alone one which unifies it with the other physical interactions."[60]

---

56   John Barrow, *The World Within the World*, 308.

57   Ibid., 310.

58   Stephen Hawking, *A Brief History of Time*,112.

59   Stephen Hawking, "Is the End in Sight of Theoretical Physics?," in *Stephen Hawking's Universe*, ed. John Boslough, 133–34.

60   Ibid., 133.

Another important area of research in view of the development of a Theory of Everything (TE) is Supersymmetry (Susy). The term "supersymmetry" refers to the mathematical transformations needed to relate particles of integer spin (bosons) to particles of half-integer spin (fermions). Supersymmetry posits that for every type of boson there exists a corresponding type of fermion with the same mass and quantum number, and vice versa, which suggests that originally they were practically the same.

> Supersymmetry demands [...] that every fundamental particle in Nature has what is called a "superpartner" with a spin that differs from that of the original particle by half a unit of spin. There needs to be a 0-spin "selectron" as partner to the electron, a 0-spin "squark" to accompany each variety of quark, a ½-spin "photino" to partner the photon, a ½-spin "wino" and "zino" as respective partners for the W and Z bosons, etc., etc.[61]

This doubling serves a practical purpose: it allows the scientists to develop mathematical constructs in which infinities are cancelled out by symmetries: "The virtual particle/antiparticle pairs of spin ½ and 3/2 would have negative energy, and so would tend to cancel out the positive energy of the spin 2, 1, and 0 virtual pairs. This would cause many of the possible infinities to cancel out."[62] On the other hand, it became evident that the known elementary particles were definitely not partners of each other under supersymmetry. So one had to make the assumption that supersymmetry relates *known* bosons and fermions to presently *unknown* fermions and bosons, as well as to explain why the "other half" have not yet been observed. For the proponents of the theory the answer to this query is simple: supersymmetry would have been manifest at high energy scales; but must have been "broken" at significantly lower temperatures. So, "super-partners" would be observed only at accelerators operating at these high energy scales.

Supersymmetry also became part of "string theory." In the 1960s this theory was still limited to the study of the working of the strong force. Its novelty consists in using 1-dimensional oscillating lines (strings), instead of the classical point particles; these vibrating lines can travel in space-time and form "world-sheets" that interact with each other by splitting and joining. Typical of strings is that they can vibrate in a specific manner, thus giving the observed particles their proper flavor, charge, mass, and

---

61   Roger Penrose, *The Road to Reality,* 875.
62   Stephen Hawking, *A Brief History of Time,* 166.

spin. In 1984, string theory incorporated supersymmetry; this enabled the string theorists to connect bosons and fermions, as well as to reduce the number of space dimensions required by the theory from 26 dimensions to 10 or 11.

Yet, in the subsequent years, string theory ran into problems: The elaboration of various versions of the equations led to the emergence of five major string theories, each of them with a varying number of curled up dimensions and differing characteristics, such as open loops and closed loops. Facing up to this difficulty, Edward Witten, the most famous proponent of string theory, in 1994 opined that the five different versions might be describing the same thing seen from various perspectives. In the words of Stephen Hawking:

> String theorists are now convinced that the five different string theories [...] are just different approximations to a more fundamental theory, each valid in different situations. That more fundamental theory is called M-Theory [....] No one seems to know what the "M" stands for, but it may be "master," "miracle" or "mystery." It seems to be all three.[63]

M-theory is hailed by its adherents as the Theory of Everything (TE). Indeed, string theory has no problem in bringing all the existing (and not yet discovered) particles into the picture; so it has no difficulty in describing the specific properties of the spin-2 particle graviton, the carrier of the gravitational force. Its theorists are convinced that they provided the most impressive step towards a theory of quantum gravity. In a 1998 interview Edward Witten declared: "String theory has the remarkable property of *predicting gravity.* [...] The fact that gravity is a consequence of string theory is one of the greatest theoretical insights ever."[64]

Not all scientists will share Witten's enthusiasm. Roger Penrose, for one, is rather skeptical about the achievements of string theory. For him, the string theorists are primarily fixated on cancelling out infinities (and thus on renormalizing their theory) with the help of refined symmetries and dualities. But this purely and highly technically mathematical concern threatens to make them lose track of the thorny problems that arise from the strict domain of physics. He writes:

---

63   Stephen Hawking and Leonard Mlodinow, *The Grand Design* (London: Bantam Press, 2010), 171.

64   Brian Greene, *The Elegant Universe*, 210 (Interview with Edward Witten, May 11, 1998).

Those who come from the side of Quantum Field Theory [string theorists being part of them] would tend to take renormalizability—or more exactly, *finiteness*—as the primary aim of a quantum-gravity union. On the other hand, we from the relativity side would take the deep conceptual conflicts between the principles of quantum mechanics and those of general relativity to be the centrally important issues that needed to be resolved, and from whose resolution we would expect to move forward to a new physics of the future.[65]

There are apparently two different approaches in the search for the holy grail of quantum gravity: one that seeks to integrate the findings of relativity into quantum physics (as this is done in string theory), and one that seeks to incorporate quantum physics into general relativity (as this is undertaken by Penrose).

The divergence between the two groups becomes manifest in their respective elaboration of space-time. String theory operates with 10 or 11 space dimensions, 6 or 7 of them being curled up at an infinitesimally small level, intertwined in such a way as to form different types of 6-holded manifolds (technically called Calabi-Yau shapes). The aim and purpose of these curled-up dimensions is to explain the immense variety of string vibrations that account for the specific properties (mass, energy, charge, spin, etc.) of the specific fermions and bosons. In this proliferation of particles, conditioned by the curled-up space dimensions, the graviton is just a specifically oscillating string among other oscillating strings; it is rather the result of a particular curled-up space formation, than that it comes to bear on the specific 'curvature of space-time, as is the case in general relativity.

Scientists, like Penrose, coming from the relativity perspective, on the contrary, focus on the role played by gravity in the curvature of space time. If this is true, then the urgent problem that needs to be tackled is how the specific geometry that is used to describe the warping of space can be brought into rapprochement with string theory's various curled-up space dimensions. An equally urgent problem that needs to be resolved is that of the clashing notions of time: instantaneous communication as apparently happens in quantum physics is at odds with the theory of relativity, according to which nothing can travel faster than the speed of light. This shows, for Penrose, that radically new ideas about the nature of time are required: "It is my opinion," he writes, "that our present picture of physical

---

65   Roger Penrose, *The Road to Reality*, 893.

reality, particularly in relation to the nature of *time,* is due for a grand shake up—even greater perhaps than that which has already been provided by present-day relativity and quantum mechanics."[66]

---

66   Roger Penrose, *The Emperor's New Mind,* 371.

CHAPTER 10

The Origin of the Universe

The Expanding Universe

When Einstein wrote his papers, it was commonly thought that our galaxy simply was the universe. In the meantime we have come to know that there are 100 to 200 billion galaxies in the universe, each of which has hundreds of billions of stars. These galaxies are grouped in clusters that are distributed evenly through space-time. The even distribution of the clusters of galaxies suggests that the universe is homogeneous (uniform in composition), and isotropic, that is, it looks the same in all directions. It is on the basis of these premises—a universe that is both homogeneous and isotropic—that Albert Einstein in 1917, shortly after the achievement of his general theory of relativity, engaged in the elaboration of equations that would depict the space-time geometry of the whole universe. Yet, when looking for a solution for his equations that would yield such a homogeneous, isotropic, and static universe, he could find none. The solutions predicted either a contracting or an expanding universe. So, in order to achieve a static universe at equilibrium, Einstein mutilated his elegant original equations by introducing a term: the so-called "cosmological constant." The cosmological constant has the same effect as vacuum energy: it acts as a repulsive, anti-gravitational force so as to balance out the attractive force of gravitation at large distances.

Einstein's apprehension was that the gravitational attraction acting on all the matter in the universe at large might eventually lead to a catastrophic contraction; so, his cosmological constant would avert this danger. Conversely, the balance obtained between gravitational attraction and the assumed repulsive force would likewise prevent the universe from expanding.

In 1922 the Russian physicist and mathematician Alexander Friedmann (1888–1925), ignoring the cosmological constant as an unnecessary "fudge factor," found a solution to Einstein's dilemma as to how to account for the fact that the universe had not already collapsed.

> Friedmann's model universe begins with zero size and expands until gravitational attraction slows it down, and eventually causes it to collapse upon itself. [Yet,] there are, it turns out, two other types of solutions to Einstein's equations that also satisfy the assumptions of Friedman's model, one corresponding to a universe in which the expansion continues forever, though it does slow a bit, and another to a universe in which the rate of expansion slows toward zero, but never quite reaches it.[1]

These three scenarios—an eventually contracting universe, a universe with infinite expansion, and a universe expanding just enough not to contract—will later be taken up by Robinson and Walker. Besides this, Friedmann reckoned with the existence of various galaxies, which in the course of the expansion of the universe would dramatically move apart with a speed proportional to their distance – in spite of the fact that within our milky way the stars remained almost stationary.

> In Friedmann's model, all the galaxies are moving directly away from each other. The situation is rather like a balloon with a number of spots painted on it being steadily blown up. As the balloon expands, the distance between any two spots increases, but there is no spot that can be said to be the center of the expansion. Moreover, the farther apart the spots are, the faster they will b e moving apart. Similarly, in Friedmann's model the speed at which any two galaxies are moving apart is proportional to the distance between them.[2]

It took some years before Friedmann's calculations reached the scientific community. In the meantime Father Georges Lemaître (1894–1966), priest-mathematician at the Catholic University Leuven in Belgium and member of the Pontifical Academy of Astronomy in Rome, had, independently of Friedmann, come to the same conclusion. In 1927 he published a paper that provided a compelling solution to Einstein's equations for the

---

1   Stephen Hawking and Leonard Mlodinow, *The Grand Design* (London: Bantam Press, 2010), 183–84.

2   Stephen Hawking, *A Brief History of Time: From the Big Bang to Black Holes* (London: Bantam Books, 1988 [1989]), 45.

case of an expanding universe; in this paper he also showed that the theory was able to explain the relation between the distances and the velocities of galaxies, just as Friedmann had done. But unlike Friedmann, who was first and foremost a mathematician, Lemaître paid attention to the physical aspects of the expanding universe in which galaxies were rushing from each other. This particular interest would occasion him in 1931 to put forward the thesis that the expanding universe must have originated from a "primeval atom" (initial singularity) that "in the beginning" exploded in a tremendous display of firework. "If you trace the history of the universe backward into the past, it gets tinier and tinier until you come upon a creation event —what we now call the big bang."[3] Some scientists suspected Lemaître of using this thesis so as to give credence to the biblical creation narrative—think of the opening verses of Genesis: "And God said, 'Let there be light,'" a saying that could be associated with the explosive eruption of an ocean of fire giving birth to the universe. It took Einstein a long time to eventually accept the reality of an expanding universe. He did so only after Edwin Hubble provided in 1929 empirical evidence for the expansion of the universe.

To situate Hubble (1889–1953) one must know something about the use he made of two important techniques that had been developed in the first half of the 19th century: spectroscopy and the measurement of the effect of a moving source of sound (the Doppler Effect). Spectroscopy is one of the astronomer's important tools for studying the properties and motions of stars. Already Newton

> discovered that if light from the sun passes through a triangular-shaped piece of glass, called a prism, it breaks up into its component colors (its spectrum) as in a rainbow [...] The different frequencies of light are what the human eye sees as different colors, with the lowest frequencies appearing at the red end of the spectrum, and the highest frequencies at the blue end.[4]

In 1815 the Munich optician Joseph Fraunhofer discovered that the light reaching us from a bright star like Sirius not only breaks up in its spectrum of colors, but that regularly dark lines run through the colors of the spectrum He came to this discovery by mounting a prism in front of the lens of a small telescope.

---

3   Stephen Hawking and Leonard Mlodinow, *The Grand Design*, 184.

4   Stephen Hawking, *A Brief History of Time*, 40–41.

The dark lines [also called absorption lines] were always found at the same colors, each corresponding to a definite wavelength of light. The same dark spectral lines were also found by Fraunhofer *in the same positions* in the spectrum of the moon and the brighter stars. It was soon realized that [...] each line is due to the absorption of light by a specific chemical element. So, it became possible to determine that the elements on the Sun, such as sodium, iron, magnesium, calcium, and chromium, are the same as those found on Earth.[5]

Spectroscopy would soon enable scientists to identify the chemistry of other stars in our galaxy, once appropriate telescopes became available. It proved to be the case that all the stars in our Milky Way possess the same chemical make-up.

A second important tool is the Doppler Effect, named after its discoverer Johann Christian Doppler (1803–1853), professor of mathematics in Prague, and specialist in the study of sound waves. Doppler found that as the moving source of sound approaches, it sounds at a higher pitch (corresponding to a higher frequency of sound waves), and when it passes and goes away, it sounds at a lower pitch (corresponding to a lower frequency). Higher frequency of sound waves means shorter wave length; lower frequency means longer wave length:

Just go out to the edge of a high way and notice that the engine of a fast automobile sounds higher pitched (i.e., a shorter wave length) when the auto is approaching than when it is going away [...] The Doppler Effect for sound waves was tested by the Dutch meteorologist Christopher Buys-Ballot in an endearing experiment in 1845—as a moving source of sound he used an orchestra of trumpeters standing in an open car of a railroad train, whizzing through the Dutch countryside near Utrecht.[6]

The bystanders along the railroad were flabbergasted: they could hardly believe their ears when witnessing the higher and lower pitched variations of well-known melodies as the train passed by.

The Doppler Effect applies also to light waves. When a source of light at a constant distant (e.g., a close star.) emits light, this emission will be captured by us at a constant frequency: "Obviously, the frequency of the waves we receive will be the same as the frequency at which they are

---

5   Steven Weinberg, *The Three First Minutes: A Modern View of the Origin of the Universe* (New York: Basic Books, Inc, 1977), 15.

6   Ibid., 13.

emitted." With a source that starts moving toward us this will be different. Indeed, "when the source emits the next wave crest it will be nearer to us, so the time that wave crest takes to reach us will be less than when the star was stationary. This means that the time between the two wave crests reaching us is smaller, and therefore the number of waves we receive each second (i.e., the frequency) is higher than when the star was stationary. Correspondingly, if the source is moving away from us, the frequency of the waves we receive will be lower."[7] If one would let these waves pass through a prism that is mounted on a strong telescope, one would see what is called "the red shift" or "the blue shift." Redshift happens when light seen coming from an object that is moving away is proportionally increased in wavelength, or shifted to the red end of the spectrum, whereas blueshift points to a decrease in wavelength when a light-emitting object moves toward the observer.

Edwin Hubble worked with one of the strongest telescopes of his time at the Mount Wilson Observatory in California. This allowed him to look into galaxies outside our Milky Way which, prior to that time, had been thought to be spiraling clouds of gas (nebulae). He was able to identify myriads of stars in them and get a picture of their chemistry. When examining the spectra of stars in faraway galaxies, he discovered something astonishing: in these spectra "there were the same characteristic sets of missing colors [indicated by the absorption lines] as for stars in our own galaxy, but they were all shifted by the same relative amount toward the red end of the spectrum."[8] This meant that the galaxies were receding from us and from each other. Hubble could even observe that the more distant a galaxy, the faster it is receding. If a galaxy located at a distance of 50 million light years is receding at a certain speed, a galaxy at twice that distance will recede at twice that speed. This is known as Hubble's law: "Hubble estimated the distance of 18 galaxies from the apparent luminosity of their brightest stars, and compared these distances with the galaxies' respective velocities, determined spectroscopically from their Doppler shifts. His conclusion was that that there is a rough linear relation between velocities and distance,"[9] as predicted by Friedmann and Lemaître: the farther apart, the faster the galaxies rush away from each other: no better proof can be given of the ongoing expansion of the universe! Upon the publication of

---

7   Stephen Hawking, *A Brief History of Time*, 41.

8   Ibid., 40.

9   Steven Weinberg, *The Three First Minutes*, 25.

Hubble's results in 1929 Einstein said that changing his equations to obtain a stable universe was "the greatest blunder of his life."

Hubble's empirical proof of the expansion of the universe aroused a new interest in Friedmann's three models of an expanding universe. In 1935 the American physicist Howard Robinson and the British mathematician Arthur Walker retrieved these models in an attempt at determining which of the three was the most probable. Yet, in this attempt they met with a serious difficulty. In order to make out whether we live in a closed or an open universe, for instance, one would have to know the average matter density of the universe, but this is a quasi-impossible enterprise, for, in addition to visible matter, one must also reckon with the presence of "dark matter":

> If we add up the masses of all the stars that we can see in our galaxy and other galaxies, the total is less than one hundredth of the amount required to halt the expansion of the universe, even for the lowest estimate of the rate of expansion. Our galaxy and other galaxies, however, must contain a large amount of "dark matter" that we cannot see directly, but which we know must be there because of the influence of its gravitational attraction on the orbits of stars in the galaxies.[10]

According to recent estimates, dark matter would constitute 84.5% of the matter of the universe. Dark matter was first discovered by the Swiss astrophysicist Fritz Zwicky of the California Institute of Technology in 1934, when studying the motions of galaxies in the Coma cluster. He pointed out that the high velocity of orbiting galaxies far away from the center of their galactic cluster could only be explained by the presence of "invisible matter" that kept them linked to the central part of the cluster.

The three models studied by Robinson and Walker have a specific graphic presentation, depending on their speed of expansion, which in turn determines whether one has to do with a finite or infinite universe.

(a) The first model is that of the *Closed Universe*. In this model "the universe is expanding sufficiently slowly that the gravitational attraction between the different galaxies causes the expansion to slow down and eventually to stop. The galaxies then start to move toward each other and the universe contracts."[11] The graphic form of this model is a semicircle whose two ends touch a horizontal line. The ascending part indicates the expansion of the universe, while the descending part indicates its contraction,

---

10   Stephen Hawking, *A Brief History of Time*, 49.
11   Ibid., 47.

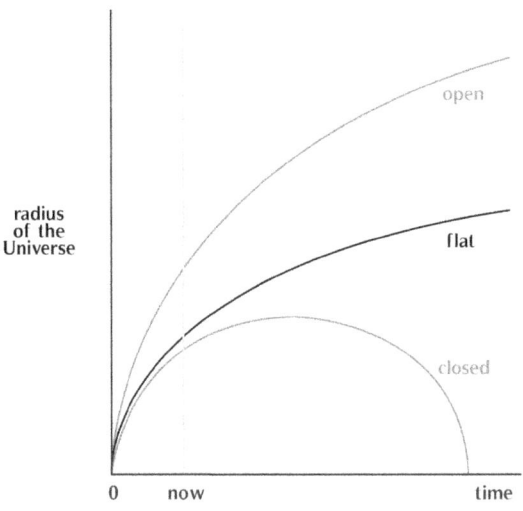

Figure 25. Three universe models

Source: David R. Gilson, *The Ever Expanding Universe in Modern Cosmology* (http://www.laidback. org/~daveg/academic/expandinguniverse/universe.PDF, p. 13), accessed September 23, 2014.

which eventually will lead to the Big Crunch. In this model the universe is finite in volume and is expected to collapse into the zero-point from which it emerged.

(b) The second model is that of the *Open Universe*. Here, the universe is conceived as "expanding so rapidly that the gravitational attraction can never stop it, though it does slow it down a bit [...] The separation between neighboring galaxies starts at zero and eventually the galaxies are moving apart at a steady speed."[12] This universe is infinite in volume and is not expected to return to its zero point. Its graphic presentation is an ascending hyperbolic arrow.

(c) The third model is that of the *Flat Universe*, also called the *Einstein-de Sitter* model. It resembles that of the Open Universe, with the only difference being that the ascending hyperbolic arrow goes down after a while, which suggests the universe's trend towards contraction. According to this model, "the universe is expanding only just fast enough to avoid recollapse. In this case the separation between the galaxies also starts at zero and increases forever. However, the speed at which the galaxies are moving apart gets smaller and smaller, although it never quite reaches zero,"[13]

---

12   Ibid.

13   Ibid.

i.e., the expansion continuously goes on, but the rate of the speed of the expansion decreases without ever reaching the zero-point. This universe, too, is infinite in volume.

Only the model of the Closed Universe predicts the final collapse of the universe under the impact of gravitational attraction. The two other models presume the emergence of sufficient "counter forces" to avoid the catastrophic implosion. It is at this juncture that renewed attention is given to the Cosmological Constant with which Einstein sought to achieve the equilibrium of the universe. Scientists began to understand that "Einstein's assumption concerning the Cosmological Constant was tantamount to assuming that a kind of repulsive, anti-gravitational force existed in the universe, and that this force could balance out gravitational attraction at large distances."[14]

The discussion of the role played by the Cosmological Constant was occasioned by the study of quantum fields, in which "vacuum energy" plays an important role. Vacuum energy results from the enormous quantities of virtual particles that are continually being created and destroyed in the quantum vacuum. Virtual particles are thought to be particle pairs that blink into existence and then annihilate each other in a time span too short to observe. Their short existence, however, is enough to help to create vacuum energy. Vacuum energy acts as a repulsive, anti-gravitational force; unlike gravity, it does not vary with the distance, since it is co-extensive with space. This explains why it is associated with the Cosmological Constant, understood as an intrinsic property of empty space.

More recently, "dark energy" (not to be confused with "dark matter") is thought to act as a Cosmological Constant. "Dark energy" is a hypothetical form of energy that permeates all of space and that accounts for 68.3% of the total mass-energy of the universe. According to the Planck mission team, the total mass-energy of the known universe contains 4.9% ordinary matter, 26.8% dark matter and 68.3% dark energy. This means that dark matter constitutes 84.5% of the *total matter* of the universe, while dark energy plus dark matter would account for 95.1% of the *total content* of the universe — only 4.9% remaining for ordinary matter. Dark energy is the most acceptable hypothesis to explain the extraordinary finding by two independent search teams[15] that the universe is expanding at an *accelerating* rate.

---

14   Richard Morris, *The Edges of Science: Crossing the Boundary from Physics to Metaphysics* (New York: Prentice Hall Press, 1990), 181.

15   The two teams are: (a) scientists from the *Supernova Cosmology Project*, Berkeley Lab U.S.A, and (b) the *High-z Supernova Search Team* in Australia.

With the help of the strong Keck telescope in Hawaii and NASA's Hubble Space telescope, the Berkeley team succeeded in detecting a "Ia supernova" that must have exploded seven billion years ago, that is, nearly at half the age of the universe. By comparing the redshift of this ancient supernova with the redshift of more nearby, younger "Ia supernovas," the Berkeley team discovered in 1998 that the universe was not just expanding at a steady rate, as had been assumed, but at an astonishingly *accelerating* rate. This finding was confirmed by the Australian team that had been independently pursuing the same search. In 2011 the directors of both teams were awarded the Nobel Prize in Physics.

The Emergence of the Universe

The tremendous expansion of the universe also raises the question as to how the universe would have looked in its primeval beginning. If the universe is still in a state of violent explosion—in which the great islands of stars known as galaxies are rushing apart at speeds approaching the speed of light—it must be possible to "extrapolate this explosion backward in time. Such a running back would show that all the galaxies must have been much closer at the same time in the past—so close, in fact, that neither galaxies nor stars nor even atoms or atomic nuclei could have had a separate existence."[16] Such an extrapolation leads us back to the era that is commonly called the early universe, close to the exploding "primeval atom" from where it all began. This is precisely the scenario that Lemaître had suggested in his account of the origin of the universe.

Standard Big Bang Theory

"Big Bang" is in the meantime the widely accepted designation for the initial fiery explosion that some 13.7 billion years ago gave rise to the formation and the expansion of the universe. Yet, it was Fred Hoyle, a vigorous opponent of Lemaître, who coined the phrase when during a BBC Radio broadcast in March 1949 he referred to Lemaître's theory as "this *big bang* idea." Hoyle was in favor of the "steady state" theory which maintained that (i) in the best Aristotelian tradition the universe did not have a beginning in time, and (ii) that as the galaxies moved away from each other, new galaxies were continuously being formed in the gaps in between from new matter that was being constantly created. The weakness of this theory was that the proposed spontaneous formation of new matter would need to include helium and hydrogen. Hydrogen and helium make

---

16   Steven Weinberg, *The Three First Minutes*, 20.

up nearly all of the ordinary matter in the universe: the most abundant element, hydrogen, accounts for 74% of the mass, while helium contributes 25%. Yet, if these elements were to be produced in the stars alone, this percentage could never be reached. In the stars, hydrogen is, to be sure, burnt into helium, and in a later stage helium is burnt into heavier elements, such as oxygen and iron. Yet, if stellar nuclear reactions were its only source of production, only a small amount of matter found in the universe should consist of helium.

Thanks to the pioneering work of George Gamow and his collaborators Ralph Alfer and Hans Bethe, we now have a satisfactory theory as to the production of the light elements hydrogen and helium in the early universe. This theory, launched in 1948 is termed the "Hot Big Bang" theory; it makes definite predictions about the universe as it exists now. It predicts (i) the creation of matter particles out of pure energy, (ii) the 3:1 ratio of hydrogen to helium, (iii) the temperature of the cosmic microwave background radiation that still permeates the universe.

(i) *The creation of matter particles out of pure energy.* At the big bang the density of the universe and the curvature of space would have been infinite. At the "singularity" of zero size, the universe is thought to have been infinitely hot. This has tremendous implications. Its first and foremost effect is the sudden proliferation of a firework of energetic radiation. At extremely high temperatures, the wavelengths of the photons, the quanta of radiation, are ultra short and thus endowed with an incredible amount of energy. The higher the temperature, the shorter the photons' wavelength and the more powerful their energy: so powerful, indeed, that—in line with Einstein's equivalence of energy and matter—elementary particles can burst forth from it. Collisions of photons with each other produce matter particles and their anti-particles out of pure energy.

The bursting forth of energy is at the same time the beginning of the explosive expansion of space. This expansion has a bearing on the gradual dropping of the early universe's extreme temperature. The rule of thumb is: "As the universe expands, any matter or radiation in it gets cooler (when the universe doubles in size, its temperature falls by half)." And further:

> Since temperature is simply a measure of the average energy—or speed—of the particles, this cooling of the universe would have a major effect on the matter in it. At very high temperatures, particles would be moving around so fast that they could escape any attraction toward each other due to nuclear or electromagnetic forces, but as they cooled off one would expect particles that attract each other

to start to clump together. Moreover, even the types of particles that exist in the universe would depend on the temperature.[17]

This would give the following scenario:

> At the big bang itself, the universe is thought to have had zero size, and so to have been infinitely hot. As the universe expanded, the temperature of the radiation decreased. One second after the big bang, it would have fallen to about ten thousand million degrees. This is about a thousand times the temperature at the centre of the Sun, but temperatures as high as this are reached in H-bomb explosions. At this time the universe would have contained mostly photons, electrons and neutrinos [...] and their antiparticles, together with some protons and neutrons.[18]

One second, thus, after the big bang, the stage was set for a "laboratory" in which the nuclei of the atoms, and much later complete atoms, would be formed together with the forces that account for their organization and mutual interactions.

(ii) *The 3:1 ratio of hydrogen to helium.* From the above description it is evident that the formation of particles, atomic nuclei, and atoms depends on the gradual dropping of the extremely high temperature, in conjunction with the rate of the early universe's expansion. Yet, in order to explain the ratio of 25% helium to 74% hydrogen another mechanism must be taken into account, namely the successive symmetry breakings that took place as the universe's temperature dropped. So a brief explanation of this mechanism is in order. I will first give the various symmetry-breakings and then come to the specific percentage of helium and hydrogen.

Some fractions of a second after the big bang the universe would have been in thermal equilibrium, with a temperature of 100,000 million degrees Kelvin ($10^{11} \, °$ K). The universe would have contained mostly photons, i.e., quanta of radiating energy that have no mass and no electrical charge. These photons gave birth, through collision with each other, to particles and their corresponding anti-particles (these emerge in pairs, with equal mass and half integral spin but with opposite electrical charges), such as electrons (and their anti-particles "positrons") and neutrinos (and their own anti-particles). When a particle and anti-particle meet, they annihilate each other in a flash of light to create new energy—i.e., new photons that through

---

17   Stephen Hawking, *A Brief History of Time,* 122.
18   Ibid., 123.

collision may dissolve again into matter-antimatter pairs, etc. At this high temperature electrons and neutrinos and their antiparticles are behaving just like so many different kinds of radiation, with an energy density that is enormous. In this way, the universe is in thermal equilibrium, i.e., it has the same uniform temperature throughout. The huge majority of the particles are leptons (electron-positron pairs and neutrino-antineutrino pairs), but besides them there are also a small number of nuclear particles (and their antiparticles): about one proton or neutron for every 1,000 million photons, electrons, or neutrinos.

This basic symmetry is broken for the first time when at a temperature of 10,000 million degrees Kelvin (10^{10} ° K), neutrinos and antineutrinos, which only weakly interact with each other and with other particles, stop annihilating and begin to behave like free particles, no longer in thermal equilibrium with the dense "soup" of photons, electrons, and positrons. Neutrinos are quasi-massless particles: they only react to the weak force and gravity. Because they practically do not interact with their environment, they are difficult to observe in the actual universe.

A second symmetry breaking occurs when the universe has a temperature of 3,000 million degrees Kelvin (3 x 10^9 ° K). At this lower temperature no new electron-positron pairs are created out of photon radiation. So, they go on annihilating each other, but without being recreated at the same rate as this was previously the case. Yet, a tiny excess of free electrons escapes annihilation; they will much later be needed to form the atoms. What exactly caused this tiny excess of electrons is still an enigma for scientists. (Equally enigmatic is the preponderance of protons over antiprotons and of neutrons over antineutrons that also must have taken place). At any rate, once the various particle-antiparticle pairs are annihilated, what is left behind are "photons plus a tiny excess of matter over antimatter that survives as pure matter after annihilation. It is this tiny excess that accounts for our existence:"[19] Ours is a universe of matter, not of antimatter.

A third symmetry breaking, nucleosynthesis, takes place between 0,01 second and 200 seconds after the big bang: "About one hundred seconds after the big bang the temperature would have fallen to one thousand million degrees Kelvin, the temperature inside the hottest stars. At this temperature protons and neutrons would no longer have sufficient energy to escape the attraction of the strong nuclear force, and would have started

---

19   Joseph Silk, *A Short History of the Universe* (New York: Scientific American Library, 1994), 89.

to combine together to produce the nuclei of atoms." [20] The first nuclei that were produced were those of deuterium (heavy hydrogen), made up of one proton and one neutron. "The deuterium nuclei then would have combined with more protons and neutrons to make helium nuclei, which contain two protons and two neutrons, and also small amounts of a couple of heavier elements, lithium and beryllium."[21] This production of helium nuclei could go on as long as enough neutrons were available to combine with protons. Once, however, the available neutrons were used up, the residual protons (some 74%) would form the nuclei of ordinary hydrogen atoms, the hydrogen nucleus consisting of only one proton.

This brings us to the 3:1 ratio of hydrogen to helium. In his book *The First Three Minutes*, Steven Weinberg gives a detailed account of the formation of this ratio. He starts from the assumption that in the period of thermal equilibrium one should have had a fifty-fifty proton-neutron ratio, and that in the course of the cooling of the universe this balance should have progressively been broken. The mechanism of this symmetry breaking is particle decay. In their interaction with neutrinos neutrons (which are slightly more massive than protons) might under the impact of the weak nuclear force decay into protons,[22] and this the more so the more the temperature dropped. So, when the temperature decreased, from 100,000 million Kelvin to 30,000 million Kelvin, the initial fifty-fifty proton-neutron ratio had shifted to 62% protons to 38% neutrons. At 10,000 million degrees Kelvin, the proton-neutron ratio had already reached 76% protons to 24% neutrons, whereas at 1,000 million degrees Kelvin, the period of nucleo-synthesis, the neutron-proton ratio stood at 86% protons to 14% neutrons.

So, it is easy to calculate how the 14% neutrons were used up to combine with protons, and how much helium nuclei resulted from this process. Suppose that out of 14.000 neutrons 12, 500 neutrons (the major part thus) are bonded to 12,500 protons so as to form 6,250 helium nuclei (recall: a helium nucleus is made up of two neutrons and two protons). Due to the helium atomic mass (about 4) this will yield a helium mass of 25,000. The remaining 73,500 protons (86,000 minus 12,500) suffice to provide as many hydrogen nuclei (with atomic mass about 1). This will yield a hydrogen mass of about 73,500. The ratio of hydrogen nuclei to helium nuclei is thus roughly 75% to 25% or 3:1.

---

20   Stephen Hawking, *A Brief History of Time*, 124.

21   Ibid.

22   See Steven Weinberg, *The Three First Minutes*, 105: "neutrino plus neutron yields electron plus proton."

(iii) *Cosmic microwave background radiation.* When the production of nuclei stopped the temperature further dropped to 300 million degrees Kelvin. The universe is filled now with photons and atomic nuclei (the majority being helium and hydrogen nuclei) plus a slew of free electrons roughly corresponding to the number of protons in the nuclei. The universe will go on expanding, but not much of interest will occur for 380,000 years. At that time the temperature will drop to the point at which electrons and nuclei can bond together to form stable atoms, whereas the removal of free electrons will make the universe transparent to radiation. Prior to this event, the universe was opaque because the freely wandering electrons prevented light (the photons) from forcefully spreading. But once the electrons were captured by their nuclei to form atoms, "a large expanse of space between each newly formed atom opened up, and photons were suddenly free to travel for great distances. In other words, matter and radiation were separated."[23] It is this radiation that reaches us now in the way of cosmic microwave background radiation. The hot big bang theory had predicted the existence and the temperature of the cosmic microwave background radiation. The discovery of it in 1964 by Arno Penzias and Robert Wilson was hailed as the confirmation of the theory. A couple of days before his death, Father Georges Lemaître got the news of the discovery. In 1978 Penzias and Wilson were awarded the Nobel Prize for Physics.

When working at Bell Labs, New Jersey, Penzias and Wilson were experimenting with a supersensitive, 6-meter (20 ft) horn antenna originally built to detect radio waves bounced off communications satellites. To measure these faint radio waves, they had to eliminate all recognizable interference from their receiver. So, they removed the effects of radar and radio broadcasting. But in spite of this, their detector picked up a very strange noise, "which was the same whichever direction the detector was pointed, so it must come from *outside* the atmosphere."[24] By repeating this observation during the day and during the night, and in various seasons, they concluded that the mysterious noise—the cosmic microwave background radiation—must come from beyond the solar system and even from beyond our galaxy. Realizing that they were catching a glimpse of the early universe, they assumed, together with cosmologists at Princeton

---

23   Brian May et al., *Bang! The Complete History of the Universe* (London: Carlton Books, 2006 [2009]), 44.

24   Stephen Hawking, *A Brief History of Time*, 44.

University, that what they had found was a tremendous blast of radiation that must have been released in the aftermath of the Big Bang.

Cosmic microwave background radiation is one of the most studied phenomena in astrophysics, because it provides us with a precious, observable relic of the early universe. No wonder that astrophysicists set out to make sure that this observed relic was authentic. The problem was that the frequency of the radiation measured by Penzias and Wilson was consistent with a temperature of 2.7 degrees Kelvin. But how could this radiation have become so cool since at the moment of its emission (when matter particles and photon radiation separated) the universe had a temperature of 3,000 degrees Kelvin? The answer, though, is quite simple: "As radiation travelled towards us, the space through which it was moving was continually expanding, stretching the light to longer and longer wave lengths, and hence leading to cooler and cooler apparent temperatures. This is our first encounter with the phenomenon known as red shift."[25] The radiation was so greatly red-shifted that it appeared to us now as cosmic microwave background radiation.

Amendments to the Big Bang Theory

New advancements in the study of quantum physics prompted scientists to slightly amend the standard Big Bang Theory. These amendments relate (i) to a deeper insight into the possible unification of the forces, (ii) to the discovery of quarks as the constituents of protons and neutrons (and of mesons), and (iii) to the plausibility of the inflation theory. The timeline prior to the period of nucleosynthesis now reads as follows.

-*Planck Epoch*: from zero to 10^{-43} seconds. This is the earliest moment in time and perhaps also the shortest possible time interval. At this point, the universe spans a region of only 10^{-35} meters (1 Planck Length), and has a temperature of over 10^{32}°C (the Planck Temperature). It is hypothesized that in the Planck epoch the four fundamental forces all have the same strength, and are possibly even unified into one fundamental force.

-*Grand Unification Epoch*, from 10^{-43} seconds to 10^{-36} seconds. The force of gravity separates from the other fundamental forces, and the earliest elementary particles (and antiparticles) begin to be formed. Compared to the other fundamental forces gravity is special in that it tends to break away from thermal equilibrium (i.e., from the process in which particles, through

---

25   Brian May et al., *Bang! The Complete History of the Universe*, 45.

mutual interaction, reach the same temperature). Rather than respecting symmetries, gravity, in the course of the cosmic expansion, evolves from a low-entropy to a high-entropy state, as will later become evident in the formation of galaxies: "We are used to thinking about entropy in terms of an ordinary gas, where having the gas concentrated in small regions represents *low* entropy, and where in the *high*-entropy state of thermal equilibrium, the gas is spread uniformly. But with gravity, things tend to be the other way about. A uniformly spread system of gravitating bodies would represent relatively *low* entropy [...], whereas *high* entropy is achieved when the gravitating bodies clump together."[26]

-*Inflationary Epoch:* from 10^{-36} seconds to 10^{-32} seconds. The inflationary epoch is a period inserted in order to resolve some apparent flaws in the standard big bang theory. The inflationary theory was launched in 1981 by Alan Guth and is now commonly accepted. Guth found out that the homogeneity and isotropy of the universe could not well be accounted for in the big bang model. The issue is namely that widely separated regions of the observable universe all have the same temperature, in spite of the fact that they are moving from each other at tremendous speeds (in tandem with space-time that is expanding faster than the speed of light). Yet it is difficult to explain this thermal equilibrium, since in the big bang scenario these regions apparently have never come into causal contact, not in the immediate past and not in earliest times. Rolling back the film of the expansion, the universe should have 1 cm diameter at 10^{-35} seconds after the big bang. Yet, "Small as this diameter is, it is still vastly larger than the distance travelled by light at that instant, about 3 x 10^{-25} centimeter."[27] So, in the standard hot big bang scenario it has not been possible for distinct small regions in the primeval universe to come into contact through light signals.

In order to get out of this deadlock, Alan Guth admitted of an extremely rapid exponential expansion that suddenly took place from 10^{-36} seconds to 10^{-32} seconds after the big bang. The cosmic inflation is thought to be triggered by the separation of the strong nuclear force from the electroweak force. In this separation the vacuum energy was created that caused the inflation to happen. "It was as if a coin 1 centimeter in diameter suddenly

---

26   Roger Penrose, *The Road to Reality: A Complete Guide to the Laws of the Universe* (London: Jonathan Cape, 2004), 706.

27   Joseph Silk, *A Short History of the Universe*, 81.

blew up to ten million times the width of the Milky Way."[28] Is it thanks to this exponential expansion, faster than the speed of light, that the regions could be separated that we now perceive in the skies to be rushing apart at tremendous speeds. Further, the fact that these separate regions all have the same temperature can be easily explained as a result of the extremely rapid inflation. Prior to this event, the universe was filled with an undifferentiated "soup" of quarks (and antiquarks) and electrons (and positrons) that were in thermal equilibrium. So, it was quite natural that, after the breath-taking exponential inflation broke up this "soup" into differentiated regions, the thermal equilibrium remained in place.

In the meantime Alan Guth's theory of the baby universe's terrific burst of expansion seems to have been confirmed. On March 17, 2014 *Scientific American* announced "that physicists have found a long-predicted twist in light from the big bang that represents the first image of ripples in the universe called gravitational waves." Gravitational waves are tremors in space-time caused by intense gravitational forces. These tremors had been predicted by Einstein; they were set in motion by the spectacular cosmic inflation. The proof of the gravitational waves, however, is indirect; it "comes in the form of a signature called B-mode polarization, a curling of the orientation, or polarization, of the light [...] left over from just after the big bang, known as the cosmic microwave background."[29] In B-mode polarization the undulations of electromagnetic waves take on particular swinging patterns. With their specialized telescopes a search team at the South Pole had reportedly discovered these swirly patterns in light waves, which would indirectly prove the existence of gravitational waves and the correctness of the theory of cosmic inflation.

-*Electroweak Epoch*: from 10^{-36} seconds to 10^{-12} seconds. The electroweak epoch began when the strong force separated from the electroweak force. At that moment, the universe was filled with hot quark–gluon plasma that began to spread over the universe's rapidly increasing volume. Through particle interactions large numbers of exotic particles were created, including W and Z bosons and Higgs bosons. At the end of this epoch W and Z bosons ceased to be created, which would occasion the symmetry breaking between the electromagnetic force and the weak force. In 2000 CERN, through the collision of lead neutrons, succeeded in giving evidence of the existence of "an exotic state of matter known as quark-gluon plasma,

---

28   Stephen Hawking and Leonard Mlodinow, *The Grand Design,* 129.
29   *Scientific American,* March 17, 2014.

in which hundreds of ordinary protons and neutrons melt together and form a fiery soup of free-roaming quarks and gluons."[30]

-*Quark-Antiquark Period*: From 10^{-32} seconds to 10^{-6} seconds. As the inflationary period ended, the universe consisted mostly of energy in the form of photons. As to the spin ½ particles (the quarks) that existed, they could not bind themselves into larger stable particles because of the enormous energy density:

> The universe was a sizzling ocean of quarks, each of which had a vast amount of energy, moving at a huge speed. [...] Whenever a quark met an antiquark in the primeval universe both would vanish, releasing a flash of radiation. The reverse process also occurred; radiation of sufficiently high energy (certainly at the energies found at this early stage of the evolution of the universe) could spontaneously produce pairs of particles, each pair composed of a particle and its antiparticle.[31]

In the description of the "firework in the beginning" the focus is now on annihilating quark-antiquark pairs rather than on annihilating electron-positron pairs as in the standard hot big bang theory. Scientists are apparently focusing on the formation of protons and neutrons out of quark triplets. Quarks are almost the size of electrons. They must have existed in larger numbers than electrons: six quarks for one electron.

-*Quark Confinement Period*:[32] from 10^{-6} seconds to 1 second. In this epoch,

> when the temperature dropped below a critical value of about 10,000 million degrees, the quarks slowed down enough to enable them to be captured by their mutual (strong force) attraction. Bunches, each of three quarks, clumped together to form our familiar protons and neutrons (collectively known as baryons), whereas the antiquarks clumped together to form antiprotons and antineutrons (antibaryons). Had the number of baryons and antibaryons been equal, the most likely outcome is that collisions between them would have resulted in complete annihilation.[...] Due to reasons we do not yet fully understand, for every billion antibaryons there were a billion *and one* baryon, so that when the grand shoot-out

---

30   Graham Collins, *Fireballs of Free Quarks*, in *Scientific American*, April 21, 2000.

31   Brian May et al., *Bang! The Complete History of the Universe* (London: Carlton Books, 2009 [2006]), 34–35.

32   Also called *Hadron Epoch*. Recall that baryons are a subcategory of the hadrons.

was over, almost all the antibaryons had vanished leaving behind the residue of protons and neutrons which make up the atomic nuclei of today.[33]

In this epoch, also quasi-massless neutrinos began to be created: protons colliding with electrons fuse to form neutrons and give off neutrinos.

-Lepton Epoch: from 1 second to 3 minutes. Once free protons and neutrons emerged, the cosmic scene was dominated by the ongoing annihilation of electron-positron pairs; this again released energy in the form of photons, whereas colliding photons in turn created more electron-positron pairs. This went on till a symmetry break took place that left over a small residue of electrons. Then, around 3 minutes after the big bang the grand event of nucleosynthesis took place, which, 380,000 years later, would lead to the creation of atoms and eventually of spiraling clouds of gas and dust from which the stars and galaxies are formed.

The Formation of Galaxies and Stars

The more refined our telescopes, the farther back we can look into the past history of the universe. Thanks to the refined Keck telescope it became possible for astronomers to get an idea of the formation of stars in early galaxies. In 2007 six star-forming galaxies were discovered about 13.2 billion light years (light travel distance) away; they must have been created 500 million years after the big bang. In 2011 astronomers were even able to spot a galaxy that was 13.22 billion years old: it must date back to 480 million years after the big bang. So, the epoch put forward for the appearance of brilliant stars is about 400 million years after the big bang. Prior to that, the universe went through a Dark Age (from 150 million to 400 million years).

Clumping of Dense Regions

A first question to be asked is on the basis of which mechanisms the "smooth" primeval universe became "lumpy," that is, filled with clusters of galaxies. The cosmic microwave background radiation discovered by Penzias and Wilson actually confronted astrophysicists with a new problem:

> The radiation appeared to be absolutely uniform; there seemed to be no variations linked with direction. [...] But the universe we see today is "lumpy"; there are huge distances between the relatively dense galaxies, which are themselves

---

33   Brian May et al., *Bang! The Complete History of the Universe*, 35–36.

grouped into clusters, and the clusters into superclusters. These superclusters are themselves separated by enormous voids, now beginning to be seen in detail in surveys such as the Anglo-Australian 2 Degree Field (2dF) Survey and the Deep Sky Survey, which reach out a billion light years from Earth. The picture of our universe emerging from these observations is certainly not uniform. [...] Hidden somewhere in the seemingly uniform early universe there must be the seeds of the structure we see today.[34]

The enigma was partially resolved when the NASA-satellite COBE (COsmic Background Explorer), launched in November 1989, detected that the Cosmic Background radiation displayed ripples, which means that "matter at the time when the Cosmic Background radiation was emitted was not absolutely uniform."[35] In 2001 a new NASA-satellite was launched, the WMAP (*Wilkinson Microwave Anisotropy Probe*) to collect more precise data of the cosmic microwave background radiation in a mission that would be spread over various years. The picture of the radiation released in 2010 "reveals temperature fluctuations—shown as color differences—dating back 13.7 billion years."[36]

Thanks to these new data, it became possible to make computer simulations of the evolving universe in which the regions with hotter fluctuations (red-colored) were marked off from those with cooler fluctuations (blue- and black-colored). Hotter fluctuations meant higher density, that is, greater accumulation of gaseous matter, whereas lower temperatures meant a lower density, and smaller accumulations of gaseous matter.

It also became clear that the force of gravity must have played an important role in the unequal accumulation process:

> The dense regions in the early universe had a greater gravitational pull than the regions that were less dense, and so drew in material from their surroundings. These, of course, further increased their gravitational pull—and so on, the process accelerating all the time. Here, as has often been the case, the rich got richer and the poor poorer! Inside each of these denser regions there were further localized variations in density, and the same sorts of processes operated—greater mass, greater pull, more runaway collapses.[37]

---

34   Ibid., 46–47.

35   Ibid., 48.

36   Stephen Hawking and Leonard Mlodinow, *The Grand Design*, 138.

37   Brian May et al., *Bang! The Complete History of the Universe*, 52.

The regions that saw their mass gradually swallowed by the increasing mass accumulation in "richer" regions were later to become the immense voids between galaxies and clusters of galaxies, whereas the dense clumps would evolve into gravitationally bound structures. Smaller systems would form first and then merge into larger conglomerations and finally coalesce into a real network of filaments, upon which, later, the first star-forming systems—small, but widening protogalaxies—would be grafted.

The force of gravity acts on matter, not only on visible matter such as clouds of gas and dust, but also on dark matter which constitutes 84.5% of the matter of the universe. It is commonly accepted that neutrinos form the bulk of the dark matter in the universe. Initially it was thought that neutrinos had zero mass. But recent measurements suggest a very small mass for, at least, one type of neutrino. Because of their tiny mass and their sheer abundance in the primeval universe, the gravitational effects of the neutrinos would have become palpable very early. Scientists opine that "the reason for the rapid formation of galactic clusters was that the neutrinos decoupled from matter a long time before the photons did. They would have formed early invisible clumps, acting as centers of mass. These centers of mass would have attracted hydrogen and helium early in the process, as soon as they were decoupled from radiation 380,000 years after the big bang."[38] In other words, the symmetry breaking through which neutrinos began to behave like free particles (see the standard hot big bang theory) had a tremendous effect on the clumping of dense regions in the primeval universe.

Appearance of the First Stars in Protogalaxies

The clumping of dense regions is the prelude to what is called the "Dark Ages" (from 150 million to 400 million years after the big bang). It is in this period that protogalaxies emerge: huge whirling clouds of dust and hydrogen gas out of which, each time, millions of protostars will be formed. Protostars are born when, due to turbulences, a whirling cloud of gas and dust breaks up into smaller parts, which then begin to collapse under their own gravity. In this collapse, the material at the center begins to heat up. A protostar looks like a star but its core is not yet hot enough for nuclear fusion to take place. Its heat and luminosity depend on the intensity of its contraction.

---

38   Armand Delsemme, *Our Cosmic Origins: From the Big Bang to the Emergence of Life and Intelligence* (Cambridge: Cambridge University Press, 1998), 30.

A protostar becomes a main sequence star when the temperature of its core exceeds 10 million K. This is the temperature needed for hydrogen fusion to operate efficiently. It takes millions of years before a protostar reaches this stage, depending on its mass: the more massive a protostar, the more rapidly it turns into a real star. So, 400 million years after the bang "the gloom was suddenly illuminated, when multitudes of stars burst forth. The universe exploded by a blaze of light."[39]

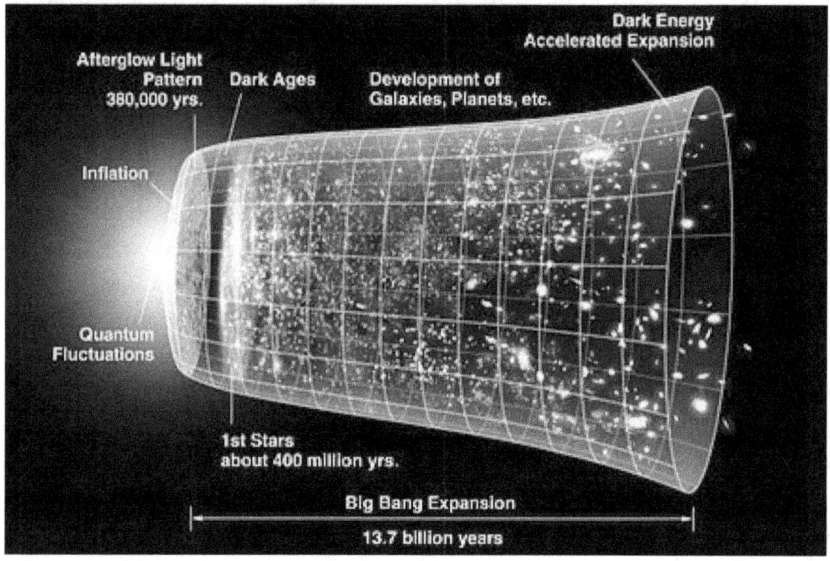

Figure 26. Chronology of the universe

Source: NASA/WMAP Science Team (http://science.nasa.gov/missions/wmap/), accessed September 23, 2014.

The source of the stars' light production is, as already mentioned, nuclear fusion, also termed astral nucleosynthesis. In this process, hydrogen nuclei are converted into helium nuclei or into nuclei of still heavier elements:

We have seen that, a hydrogen atom, the simplest of all, has a single proton as its nucleus and one orbiting electron. Inside a star the heat is so intense that the electron is stripped away from the nucleus, leaving the atom incomplete; the atom is said to be "ionized." At the star's core, where the pressure as well as the

---

39   Brian May et al., *Bang! The Complete History of the Universe*, 52.

temperature is so extreme, these nuclei are moving at such enormous speed that when they collide, nuclear reactions are able to take place. Nuclei of hydrogen are combining to build up nuclei of the second lightest element, helium.[40]

As a result of this nuclear fusion the star begins to shine and give off a tremendous amount of energy within an environment that, given the ongoing expansion of the universe, gets cooler and cooler.

The speed at which a star burns its fuel—in this case: its hydrogen nuclei—determines how long it will live. The first stars were too massive to last long: They were 30 to 100 times as massive as our Sun and millions of times as bright, burning only for a few million years before exploding as supernovas. This occurred when the universe was almost a billion years old. Given their enormous reservoir of fuel, one would have expected these stars to shine much longer than our much smaller Sun. This, however, is not the case: the giant stars burned their fuel at a faster pace, as if they were eager to stay as bright as possible for their relatively short (cosmic) lifetimes. This "ambition" predestined them to become full-fledged supernovas when they exploded: When the stars have used up their fuel, "the central regions of the star would collapse to a very dense state, such as a neutron star or black hole. The outer regions of star may sometimes get blown off in a tremendous explosion called supernova, which would outshine all the other stars in its galaxy."[41]

Supernova explosions still occur in the universe. In fact, stars eight times as massive as our Sun, and which were formed billions of years after the first generation of stars, are destined to die in that way. Almost 25 to 50 supernovas are discovered in other galaxies each year. Supernovas have also occurred in our galaxy in the past. In 1572 Tycho Brahe discovered a supernova in our Milky Way, as did Galileo in 1604.

Second and Third Generation Stars

Second generation stars, also termed "Population II stars," are thought to be born about 3 billion years after the big bang, followed by the birth of third generation stars, also termed "Population I stars," about 6 billion years after the big bang. It is possible to locate them by paying attention to two major types of galaxies: (a) elliptical galaxies, with a round shape, and (b) spiral galaxies, which behave like rotating disks. The type of galaxy formed depends on the initial rate of star production in it. If the new galaxy forms

---

40   Ibid., 54.

41   Stephen Hawking, *A Brief History of Time*, 126.

stars slowly, then the gas cloud has enough time to elongate into a flat disk-shaped spiral; if the galaxy forms stars quickly, it uses up its supply of gas before succeeding in forming a disk; it remains elliptic.

Elliptical galaxies. All of them had a quick and high rate of star production, which eventually exhausted their gas supply. So, they no longer have active star formation and mainly consist of old, second generation stars. Unlike the spiral galaxies in which the stars have ordered rotations, the stars of elliptic galaxies are on orbits that are randomly oriented. All elliptical galaxies probed so far have supermassive black holes in their center.

Spiral galaxies. They formed more slowly and still have ongoing star formation today. They are very thin and rotate rapidly in circular orbits. Unlike elliptical galaxies, all their stars move in the same direction, which leads to the formation of spirals. The scenario of their origin reads as follows: Spiral galaxies are composed mostly of gas and dark matter. As a galaxy gains mass (by absorbing smaller galaxies) the dark matter component stays on the outer parts of the galaxy, whereas the inner gas component quickly contracts and begins to rotate; this results in a very thin but rapidly rotating disk. This disk rotates around a central concentration of

Figure 27. Spiral galaxy

Source: ESA/Hubble (http:www.spacetelescope.org/images), accessed September 23, 2014.

older stars known as the bulge (kept together by a supermassive black hole) and is surrounded by spiral arms that contain a great many young, blue stars that are still being created. Of the pictures shot with the help of potent telescopes spiral galaxies are the most impressive. They make up approximately 60% of the galaxies in the local universe.

It is relatively easy for astronomers to tell the difference between second and third generation stars: what distinguishes both types is the percentage of heavy elements they contain. Take our Milky Way: Despite the fact that second generation stars are few in number in our Milky Way,

> they have a unique attribute that enables astronomers to recognize them. At the beginning of the galaxy, very few heavy elements had been produced. Consequently, the first stars would have had essentially no oxygen or iron. As time went on, supernovas ejected debris into the interstellar gas, and stars continued to form out of this polluted environment. Thus, the more recently a star formed, the higher is its content of metals relative to hydrogen. In fact, stars that formed relatively recently, for example, 10 million years ago, are about twice as metal-rich as the Sun. Conversely, stars that formed before the formation of the solar system, some 4.7 billion years ago, are metal-poor in comparison with the Sun. The oldest stars in the galaxy have only one-hundredth or one-thousandth of a percent of the metal content of the Sun, as measured spectroscopically.[42]

These metal-poor stars are second generation stars, the oldest components of disc galaxies like ours.

As already stated, the source of a star's light production is astral nucleosynthesis. In the first generation stars hydrogen nuclei burned into helium nuclei, made up of two protons and two neutrons. Yet, the nucleosynthesis that takes place in the next generations of stars goes much further than the making of helium; it will lead to nuclear fusions resulting in heavier elements, such as carbon (six protons and six neutrons), oxygen (eight protons and eight neutrons), and iron (26 protons and 30 neutrons). In fact, nuclear fusions are a question of life and death for the stars, especially for massive stars, i.e., stars of at least eight solar masses. These feel the urge to burn heavier elements in order to survive. The basis of this mechanism is easy to understand: in order for a star to keep up its mechanical equilibrium, it must produce sufficient energy to buffer itself against the pull of gravity that threatens it with inward implosion. For a star like our Sun, the conversion of hydrogen into helium suffices to create this

---

42   Joseph Silk, *A Short History of the Universe,* 222.

energy. But massive stars, after having almost used up their reservoir of hydrogen, have no other choice than to fuse their helium nuclei with additional protons and neutrons so as to convert them into heavier elements: lithium, beryllium, carbon, oxygen, and iron.

The more the massive stars are forced to burn heavier elements, the more they are condemned to eventually end up in a catastrophic disruption: their central parts collapse to form white dwarfs, neutron stars, or black holes, while their outer regions get blown off in a supernova explosion, during which elements heavier than those already produced by nucleosynthesis are being formed. Yet, such a disruption indirectly contributes to the emergence of a next generation of metal-rich stars. Reflecting on the birth of our solar system, Stephen Hawking has this to say:

> Some of the heavier elements produced near the end of the star's life would be flung back into the gas in its galaxy, and would provide some of the raw material for the next generation of stars. Our own Sun contains about 2 percent of these heavier elements because it is a second- or third-generation star, formed some five thousand million years ago out of a cloud of rotating gas containing the debris of earlier supernovas. Most of the gas in that cloud went to form the Sun or got blown away, but a small amount of the heavier elements collected together to form the bodies that now orbit the sun as planets like the Earth.[43]

Our Sun and its planetary system immensely benefited from the "cosmic catastrophe." Indeed, biological life on Earth could not have emerged if massive stars had not previously produced carbon, oxygen, and iron, as essential ingredients of life.

---

43   Stephen Hawking, *A Brief History of Time*, 126.

CHAPTER 11

The Place of Humans in the Universe

Anthropic Principle

As we saw in the preceding chapter, the essential ingredients of life—carbon, oxygen, and iron—were sacrificial gifts bequeathed to us as a result of the death of massive stars that no longer exist. This awareness leads us to a consideration of the anthropic principle, which postulates that in some regions of the universe, the physical conditions must have become such that they paved the way for the emergence of beings like us, able to reconstruct, on the basis of their developed intelligence, their past cosmic history. Among these favorable conditions are what happened in the stars and interstellar space. Indeed, for carbon-based life to appear, elements heavier than helium and hydrogen are needed. Through stellar nuclear fusions, massive stars produced carbon, oxygen, and iron indispensable for the emergence of biological life. When the second-generation stars exploded as supernovae, they scattered these life-generating elements into space, an explosion that led to the formation of still heavier elements such as zinc and iodine (iodine being essential to the healthy functioning of the thyroid gland).

In reflecting on these facts, John Polkinghorne, Anglican priest and mathematical physicist, calls attention to a whole range of delicate balances that seem to have been orchestrated by nature in the evolution of the cosmos, such as the fine-tuning that played an important role in the burning of heavier elements in the stars: "To make carbon in a star, three helium nuclei have to be made to stick together. This is tricky to achieve and only possible because a special effect (technically called a resonance) is present in just the right place. This delicate positioning depends upon the strong nuclear force that holds the nuclei together. Change this force a little, and you lose the resonant effect."[1] The same special effect is needed to produce oxygen (by making another helium nucleus stick to the carbon nucleus). The list of this amazing fine-tuning can easily be extended. For stars to burn uniformly for a long period of time a delicate balance between gravity and the electromagnetic force is needed. If this balance is disturbed, the stars either cool too quickly to act as energy sources, or become so hot that they burn away after only a few million years.

A further reason for amazement is that both the age and the size of the universe concur in the process. Six billion years of cosmic expansion were needed for carbon-producing stars to explode as supernovas. And given the correlation between the age and size of the universe, a universe approximately the size of ours—with its hundreds of billions of galaxies—would have been required in order to make possible the emergence of carbon-based, intelligent life on earth. The solar system is thought to have been formed 9 billion years after the big bang, whereas the first forms of life appeared on Earth 10 billion years after the big bang (or 3.7 billion years ago), in the form of primitive aquatic organisms much like modern-day bacteria. It took another 2 to 3 billion years before an atmosphere conducive to the evolution of land-based plant and animal life was formed.

Polkinghorne is not the only scientist to marvel at the astonishing fine-tuning of so many physical processes. Stephen Hawking, who cannot be suspected of giving a special place to humans in the universe, also appears to be struck by these delicate balances. "The laws of science, as we know them at present," he wrote,

> contain many fundamental numbers, like the size of the electric charge of the electron and the ratio of the masses of the proton and the electron [...] The remarkable fact is that the values of these numbers seem to have been very finely adjusted to make possible the development of life. For example, if the electric charge of the electron had been only slightly different, stars either would have been unable to burn hydrogen and helium, or else they would not have exploded.[2]

This interest in the fine-tuning brings us to the heart of what is at stake in the anthropic principle: the place of humans in the universe. The anthropic principle was first presented by Brandon Carter at a 1973 Krakow symposium honoring Copernicus' 500th birthday. In a seminal paper, Carter

---

1   John Polkinghorne, *Quarks, Chaos, and Christianity* (London: Triangle, 1994), 29.

2   Stephen Hawking, *A Brief History of Time: From the Big Bang to Black Holes* (London: Bantam Books, 1988 [1989]), 131–32.

distanced himself from the so-called Copernican principle, according to which humans do not occupy a privileged place in the universe. For him, our situation in the universe is not necessarily *central,* but it is nevertheless privileged to a certain extent. This privileged position is expressed in the anthropic principle, which states that the fundamental constants of nature must be such that "what we expect to observe must be restricted to the conditions necessary for our presence as observers."[3] In other words, we live in a very special region of the universe, one in which life and intelligence are possible. So, we are able to look back for the steps in the formation of the cosmos that paved the way for the reality of our existence. This does not necessarily reflect an anthropocentric perspective. It rather underlines the need for us to realize that we belong to a cosmos that has the power to "generate" us. There exists a deep connection between how the chemical elements came to be processed in the course of the formation of the cosmos and our carbon-based, intelligent life.

The anthropic principle has been developed to include both a "weak" and a "strong" version. The foregoing explanation is termed the "weak anthropic principle":

> We see the universe the way it is because we exist [...] One example of the use of the weak anthropic principle is to "explain" why the big bang occurred about ten thousand million years ago—it takes about that long for intelligent beings to evolve. As explained above, an early generation of stars first had to form. These stars converted some of the original hydrogen and helium into elements like carbon and oxygen, out of which we are made. The stars then exploded as supernovas, and their debris went to form other stars and planets, among them those of our Solar system, which is about five thousand million years old. The first one or two thousand million years of the earth's existence were too hot for the development of anything complicated. The remaining three thousand million years or so have been taken up by the slow process of biological evolution, which has led from the simplest organisms to beings that are capable of measuring time back to the big bang.[4]

---

3   Brandon Carter, "Large Number Coincidences and the Anthropic Principle in Cosmology," in *Confrontation of Cosmological Theories with Observation,* ed. Malcolm Longair (Dordrecht, 1974), 294.

4   Stephen Hawking, *A Brief History of Time,* 131.

So, the focus is on the right conditions that were necessary for the emergence of intelligent life on earth. Had these conditions not been just right, we would not have been here to observe them.

The weak anthropic principle limits its considerations to our locale as human beings in a particular region in the universe, and at a particular time—without moving beyond this *local perspective*. It states that our location in the universe is necessarily privileged to the extent of being compatible with our existence as observers. Yet, we can also talk about a strong anthropic principle. In this version, the anthropic principle is extended to include the *whole universe* as the possible birth place of intelligent life under certain conditions. The strong anthropic principle posits that the universe as such has a biocentric structure: "The universe (and hence the fundamental parameters on which it depends) must be such as to admit the creation of observers within it at a certain stage."[5] Contrary to what one might expect, the strong anthropic principle questions the privileged position of the humans in the whole of the universe, since it accepts that certain forms of life could also emerge in other parts of it:

> According to this theory, there are either many different universes or many different regions of a single universe, each with its own initial configuration and, perhaps, with its own set of laws of science. In most of these universes the conditions would not be right for the development of complicated organisms; only in the few universes that are like ours would intelligent beings develop and ask the question: "Why is the universe the way we see it? The answer is then simple: If it had been different, we would not be here!"[6]

In actual fact, the strong anthropic principle reckons with the possibility of the existence of multiple universes (multiverses). In his book *The Grand Design*, co-edited with Mlodinow in 2010, Hawking retrieved the scenario of the birth of multiple universes developed in 1983 by the Russian astrophysicist Andrei Linde. This scenario has been further embellished as a result of "M-theory," the unifying matrix of the various string theories, which supports the notion of the existence of a tremendously high number of parallel universes. According to this scenario, the universe at its inflationary expansion would have branched off into various "bubbles" that gave birth to additional bubbles that in turn gave rise to additional bubbles, etc.

---

5   Brandon Carter, "Large Number Coincidences and the Anthropic Principle in Cosmology," 294.

6   Stephen Hawking, *A Brief History of Time*, 131.

—which bubbles would, in various degrees, become candidates for evolving into full-fledged universes. Some of these mini-universes would expand but collapse again; others would grow large enough so that they could develop galaxies and stars; in some of them the conditions would be right to give rise to complicated life forms and, eventually, intelligent life. From the outside these bubbles would look extremely small, yet they would harbor cosmic regions as big as our universe, which is also one of these bubbles.

What sort of life, if any, might have evolved within the other bubbles cannot be verified by us. Nonetheless theoretical models predict that they can all develop their own physical laws and parameters: "In some universes electrons have the weight of golf balls and the force of gravity is stronger than that of magnetism."[7] These differences are not determined by logic or physical principles; they flow from the specific histories and quantum mechanical processes that occurred in them: "The parameters are free to take on many values and the laws to take on any form that leads to a self-consistent mathematical theory, and they do take on different values and different forms in different universes."[8] At this juncture Hawking even advances the idea of the creation of multiple universes out of nothing: "Quantum fluctuations lead to the creation of tiny universes out of nothing. A few of these reach a critical size, then expand in an inflationary manner, forming galaxies, stars and, in at least one case, beings like us."[9]

With the perspective of multiple universes Hawking takes us a step further in the de-centering of humans inaugurated by the Copernican turn. "Today we know," he writes, "there are hundreds of billions of stars in our galaxy, a large percentage of them with planetary systems, and hundreds of billions of galaxies. [We also know] that our universe itself is also one of many, and that its apparent laws are not uniquely determined."[10] The sheer number of universes (ten to the power of five hundred according to M-theory), each of them with their different physical laws, makes it highly improbable that only in our universe would the right parameters and fine-tunings have occurred for intelligent life to appear. Also in other universes the same occurrence could randomly and contingently come about, just as it randomly and contingently took place in our universe. Given enough universes, sooner or later one of them is bound, by mere

---

7   Stephen Hawking and Leonard Mlodinow, *The Grand Design* (London: Bantam Press, 2010), 142.

8   Ibid., 143.

9   Ibid., 137.

10   Ibid., 143.

chance alone, to create the particular set of conditions that favor life; and this outcome is likely to repeat itself again and again. It is like the rolling of dice. If you try long enough the same combination of sets of numbers will come out again. Probability calculus takes it for granted that in spite of so many misses a score of good luck must inevitably arrive. In a lottery there is always a winner; but it need not necessarily be you: "Those living beings [who have made it], while having cause to thank their luck, could seem to have little ground for astonishment."[11]

Chance and good luck: it would seem that this is all that is needed to produce intelligent life in one of the possible universes. We might call ourselves fortunate "in the relationship of our sun's mass to our distance from it. [..] If our sun were just 20 percent less or more massive, the earth would be colder than present-day Mars or hotter than present-day Venus,"[12] and thus not conducive to the emergence of life. Yet, Hawking goes on, it is totally unlikely that such a happy coincidence would have happened only once. Astrophysicists reckon with the fact that soon we will discover outside the solar system stars with planetary systems in which at least one planet has the right conditions under which life could develop. Equally unlikely is it that the fine-tunings in the laws of nature would have happened only once—to specifically produce us, human beings, as it were, by divine design. "In the same way that the environmental coincidences of our solar system were rendered unremarkable by the realization that billions of such systems exist, the fine-tunings in the laws of nature can be explained by the existence of multiple universes."[13] For Hawking, it is evident that the more one realizes the existence of multiple universes, the more also one is forced to admit that the fine-tunings that developed in our universe are not so exceptional and unique that we should ascribe them to an intelligent design. He writes:

> Many people through the ages have attributed to God the beauty and complexity of nature that in their time seemed to have no scientific explanation. But just as Darwin and Wallace explained how the apparently miraculous design of living forms could appear without intervention by a supreme being, the multiverse concept can explain the fine-tuning of physical law without the need for a benevolent creator who made the universe for our benefit.[14]

11   John Leslie, *Universes* (London/New York: Routledge, 1989), 15.
12   Stephen Hawking and Leonard Mlodinow, *The Grand Design*, 152.
13   Ibid., 165.
14   Ibid.

Intelligent Design Revisited

The term "intelligent design" acquired a negative reputation in scientific circles because of the use and abuse made of it by the American Discovery Institute, a politically conservative think tank that is home to an anti-Darwin lobby. The Seattle-based Discovery Institute was founded in 1990 with the purpose of promulgating a scientific underpinning for creationism so as to discredit Darwin's theory of evolution; it holds that certain remarkable features of the universe and of the living beings are best explained by an intelligent cause (an intelligent designer), and not by a random and contingent process such as natural selection. Its advocates attempt to introduce their theory in public school biology curricula, with the aim and purpose to combat the "atheistic" connotations of Darwinism. So, they lobby for having the theory of evolution replaced with the doctrine of intelligent design in the school curricula, or at least to achieve that the two *theories* be taught side-by-side as equally valid. The movement is very influential in the USA; due to its solid funding it is able to attract interested scientists for the purpose of publishing books that seek to demonstrate alleged weaknesses and lacunas in Darwin's theory, such as the missing links in fossil records, or the improbable fact that irreducibly complex systems could have emerged from successive modifications of a precursor system.[15]

"Intelligent design" has of old been the teleological argument for proving the existence of God from the good order of creation. One ought only to think of Newton who in his *Opticks* referred to God's design when asking the question as to how it came to be that the bodies of animals were contrived with so much art: for him, the eye could only have been made by somebody having the skill of optics. As an alternative to evolutionary biology, however, this argument is somewhat obsolete. It fails to pay attention to what is crucial in this theory: that complex organisms evolved from less complex organisms through contingent processes of natural selection:

> Much has been learned about natural selection during the past two centuries, and indeed there are many complicating factors involved. However, the fundamental contribution of Darwin's *On the Origin of Species* remains essential to evolution: heritable genetic variation and differential survival over the course of many

---

15   See Michael Behe, *Darwin's Black Box: The Biochemical Challenge to Evolution* (New York: The Free Press, 1996), 39.

generations can lead, eventually, to significant changes in biological population. That is the Darwin-Wallace theory of descent with modification, or natural selection.[16]

The concrete ways in which the amazing biodiversity originated on earth seems to defy any preconceived design, since many contingent factors are involved in this age-long proliferation, such as the species' urge for survival in changing biotopes, or their need to protect themselves against predators.

Contingency, however, is not consistent with well-ordered planning. So, fears arise among believers that the Darwin-Wallace theory of natural selection may lead to a world view in which the whole of biological life—human life included—is explained as the product of a series of purely random processes. It is against this background that one has to understand the endeavors of the advocates of Intelligent Design to prove that God is the sole agent in the process of creation—the intelligent cause who gives a clear direction to the development of the cosmos and of life. Yet, this endeavor lands us in a deadly antagonism: the more the advocates of "intelligent design" underline the necessity of an intelligent cause, the more atheistic scientists place the accent on chance and randomness, as we have seen with Stephen Hawking.

In order to get out of this deadlock Robert Asher, a paleontologist who himself has religious convictions, reminds both the advocates of "intelligent design" and the evolutionists of the fact that the God idea they embrace or reject is entirely one-sided. As if the creator God would not be able to call forth a self-evolving creation. From his study of fossils Asher has come to realize that the amazing biodiversity on earth cannot possibly be the result of some preconceived divine plan. In order for this biodiversity to emerge contingent factors must have come into play such as the obstacles the species had to overcome in order to survive. Time and again their organisms had to adapt to changing biotopes; they often had to develop refined characteristics (stronger wings or shells) to make them withstand attacks of predators. Only in retrospect may these biological adaptations display some directionality; but it would be nonsense, Asher maintains, to connect this "directionality" to some preconceived plan. One has rather to do with a type of "order" born from the apparently random working of things.

In short, Asher asks the question as to why we think God must act in the way we think he should act. He invites both parties to examine the extent to which their (embraced or rejected) notion of design has an

---

16   Robert Asher, *Evolution and Belief: Confessions of a Religious Paleontologist* (Cambridge: Cambridge University Press, 2012), 4.

anthropocentric ring about it: Who are we to impose our notion of design on the ineffable dealings of the Creator? He wrote:

> The equations "purpose = God" and "randomness = atheism" [have a common origin. They] are contingent upon some preconceived notion of order [...] The fact that we find it difficult to appreciate the creative power of apparent randomness does not mean that God suffers from this problem [...] Requiring that God's style of invention has to resemble our own, as do many creationists and atheists alike, seems extraordinarily presumptuous and vain. Darwinism must be God's method of action.[17]

Moreover, Darwinian evolution is, strictly speaking, not a random process; it is steered by a stubborn will to survive. The novel biological formations are indicative of a struggle for life that generates its own creative power.

Evolution of the Cosmos

The discussion about intelligent design is apparently not limited to the origin of the species; it comprises also the evolution of the whole cosmos. This becomes evident from the controversy between the American Jesuit George Coyne and Cardinal Christoph Schönborn of Vienna. During a tour in the United States Schönborn had published a column in the *New York Times* of July 7, 2005. In this column Schönborn sided with the American advocates of "intelligent design" and claimed that neo-Darwinism and the theory of multiple universes are incompatible with the church's belief in God's purposeful action. I quote from this column:

> In the 19th century, the First Vatican Council taught a world newly enthralled by the "death of God" that by the use of reason alone mankind could come to know the reality of the Uncaused Cause, the First Mover, the God of the philosophers. Now at the beginning of the 21st century, faced with scientific claims like neo-Darwinism and the multiverse hypothesis in cosmology invented to avoid the overwhelming evidence for purpose and design found in modern science, the Catholic Church will again defend human reason by proclaiming that the immanent design evident in nature is real. Scientific theories that try to explain away the appearance

---

17   Ibid., 13–14.

of design as the result of "chance and necessity" are not scientific at all, but, as John-Paul II put it, an abdication of human intelligence.[18]

The aim of the cardinal was apparently to remove any doubt about the existence of God. But Coyne, then the director of the Vatican Observatory in Castel Gandolfo, did not take it that Schönborn denied cosmic evolution. He rather saw in the evidence of cosmic evolution a welcome opportunity for Christians to re-imagine the Creator God's relationship to the world, and so to deepen their faith. Coyne's indignation is understandable if one realizes the extent to which Schönborn's reference to the First Vatican Council is in a sense obsolete (it took place from 1869 till 1870: some sixty years before the spreading of the Big Bang Theory) and how selective is his recourse to modern science (which he apparently limits to what Newton had to say about "intelligent design").

In his response Coyne does not dwell on the theory of the multiple universes. He immediately turns to cosmic evolution, which—comparable to Darwin's vision of growth in complexity—led to the formation of our solar system and to the production (in the stars) of the chemical elements needed for the emergence of life, and finally intelligent life on earth. He was confident that his view on cosmic evolution matched with recent papal teaching. He right from the start mentions that Pope John-Paul II, in an epoch-making declaration to the *Pontifical Academy of Science* in 1996, had stated that "evolution is not a mere hypothesis"; and remarks that, in the same allocution, the Pope had invited theologians "to draw reasonable implications for religious belief from that conclusion."[19] In line with this, Coyne will give his faith-inspired reading of cosmic evolution, a reading that is based on scientific facts. Coyne takes his lead from Schönborn's derogatory statement about the theory of evolution, namely that it tries to explain away the appearance of design as the result of "chance and necessity." He stresses that it is nonsense to pit chance against necessity and vice versa. What really matters is the enormous "fertility" of the universe—his term for the intrinsic creativity of the cosmos. It is only from this background that one is able to appreciate the subtle interplay between chance

---

18   Christoph Schönborn, "Finding Design in Nature," *New York Times*, July 7, 2005: http://www.nytimes.com/2005/07/07/opinion/07schonborn.html (accessed December 17, 2012). Partially quoted in Stephen Hawking and Leonard Mlodinow, *The Grand Design*, 225.

19   George Coyne, "God's Chance Creation," *The Tablet*, August 6, 2005: http://www.thetablet.co.uk/article/1027 (accessed June 20, 2012).

(stroke of good luck) and necessity (the working of the laws of nature). In Coyne's words:

> The classical question as to whether the human being came about by chance, and so has no need of God, or by necessity, and so through the action of a designer God, is no longer valid. And so any attempt to answer it is doomed to failure. The fertility of the universe, now well established by science, is an essential ingredient, and the meaning of chance and necessity must be seen in light of that fertility. Chance processes and necessary processes are continuously interacting in a universe that is 13.7 x 1 billion years old and contains trillions and trillions of stars. Those stars as they "live" and "die" release to the universe the chemical abundance of the elements necessary for life. In their thermonuclear furnaces stars convert the lighter elements [hydrogen and helium] into the heavier elements [carbon, oxygen, iron]. There is no other way, for instance, to have the abundance of carbon necessary to make a toenail than through the thermonuclear processes in stars. We are all literally born of stardust.[20]

To illustrate how chance and necessity work in tandem Coyne gives the example of how a hydrogen molecule forms:

> two hydrogen atoms meet in the early universe. By necessity (the laws of chemical combination) they are destined to become a hydrogen molecule. But by chance the temperature and pressure conditions at that moment are not correct for them to combine. And so they wander through the universe until they finally do combine. And there are trillions and trillions of such atoms doing the same thing. Of course, by the interaction of chance and necessity, many hydrogen molecules are formed and eventually many of them combine with oxygen to make water, and so on, until we have very complex molecules and eventually the most complicated organism that science knows: the human brain.[21]

The same interaction of chance and necessity was also at work in the formation of the atoms. It took almost 380,000 years after the Big Bang before the atomic nuclei were able to capture their electrons, for only by then the universe had expanded enough for its temperature to drop so as to make the combination possible.

The interaction of chance and necessity is apparently spread over billions of years starting from the Big Bang 13.7 billion years ago. It led to

---

20 Ibid.
21 Ibid.

a growing complexity of chemical elements, which in turn gave a certain intrinsic natural directionality to the evolutionary process. The more complex an organism becomes the more determined is its future. "This does not necessarily mean, however," Coyne continues, "that there need be a person directing the process, nor that the process is necessarily an 'unguided, unplanned process of random variation and natural selection' as Cardinal Schönborn describes it. It is precisely the fertility of the universe and the interaction of chance and necessity in that universe which are responsible for the directionality. Thus far the story of science."[22] The next question then to be asked is: where does the Creator God feature in this scientific scenario? Coyne's answer is simple: "If one believes in God's loving relationship with his creation, and especially with the human beings made in his image and likeness, and if one also respects the science described above, then there are marvelous opportunities to renew one's faith in God's relationship to his creation."[23]

Although Coyne nowhere mentions this, he appropriates in fact an insight developed by Alfred North Whitehead, the founder of Process Philosophy. In the closing pages of his 1929 book *Process and Reality,* Whitehead distances himself from what he calls the unilaterally ruling God. He rather prefers to speak about the "Galilean God of endearment." He wrote: "When the Western world accepted Christianity, Caesar conquered; and the received text of Western theology was edited by his lawyers. [...] The Church gave unto God the attributes which exclusively belonged to Caesar." Contrasting with this view is the Galilean notion of God. "It does not emphasize the ruling Caesar, or the ruthless moralist, or the unmoved mover. It dwells upon the tender elements in the world which slowly and in quietness operate by love. [...] Love neither rules, nor is it unmoved."[24] In a similar vein Coyne brings up the idea of a tender God. From what we now know about the evolution of the cosmos, he says, "we should move away from the notion of a dictator God, or a designer God, a Newtonian God who made the universe as a watch that ticks along regularly. Perhaps God should be seen more as a parent or as one who speaks encouraging and sustaining words. Scripture is very rich in these thoughts."[25]

In less poetic terms: we ought to imagine God as the One who is so generous that he allows the creation to share in his own creative

---

22   Ibid.

23   Ibid.

24   Alfred North Whitehead, *Process and Reality* (New York: Free Press, 1929 [1969]), 404.

25   George Coyne, "God's Chance Creation," *The Tablet,* August 6, 2005.

inventiveness. This is exactly what Process Philosophy propounds. According to this philosophy creativity is, to be sure, first and foremost exemplified in God, but as soon as God calls forth the universe, this universe in turn develops a stupendous creativity that causes ever new things to emerge. In Whitehead's approach, the whole focus is on the principle of creativity. In it, the Creator God is, to be sure, the highest creative instance, but not the sole creative instance; he allows the cosmos to develop its own intrinsic creativity. So, too, is it for Coyne: "God in his infinite freedom," he writes, "continuously creates a world that reflects that freedom at all levels of the evolutionary process to greater and greater complexity. God lets the world be what it will be in its continuous evolution. He is not continually intervening, but rather allows, participates, loves."[26] This implies that God takes the risk of letting the creative process take its own course because of the many chance processes involved in it. We should never forget that we, human beings, are the outcome of so many improbable cosmic events that took place billions and billions of years ago. We are late-comers in an evolutionary process that is so extraordinary that it fills us with feelings of awe.

Lessons from the Past

Science is able to give a detailed description of the formation, in the stars, of the chemical elements that have led to the evolution of biological life on earth, with the human brain as its apogee. But when it comes to assigning a place, in this process, to the Creator God, modern science is strangely silent. Having eliminated the alleged "God of the gaps" scientists would, for the most part, quote some truncated reminiscences from ancient beliefs. They would still use Christian formulas like "creation out of nothing" without mentioning the agent who performed the miracle. For a great many of them, the universe is "the ultimate free lunch" (Alan Guth), or as Stephen Hawking put it: "Spontaneous creation is the reason there is something rather than nothing, why the universe exists, why we exist. It is not necessary to invoke God to light the blue touch paper and set the universe going." [27]

Hawking should have omitted this last sentence, for contemporary science neither supports nor denies the existence of a deity behind the creative processes of the universe. If believers confess the existence of such a deity, they do so on the basis of a different type of rationality. An example

---

26   Ibid.

27   Stephen Hawking and Leonard Mlodinow, *The Grand Design*, 180.

in case is the paleontologist Robert Asher who terminated his book on *Evolution and Belief* with the following statement:

> Speaking for myself, I accept this deity [behind the natural processes] based on my own intuition, and I make no pretention that this intuition has any scientific basis. However, although I acknowledge my belief to be non-scientific, it is entirely rational. Science is a subset of rationality; the former has a narrower scope than the latter. To ignore rationality when it does fall beyond the scientific enterprise would be an injustice both to reason and to humanity.[28]

It is this broader type of rationality that was at work in the pre-scientific creation narratives I presented in the beginning chapters of this book. People today can still learn some lessons from them. By way of conclusion I will highlight some elements of these narratives that may provide an inspiration to believers who have assimilated the contemporary concept of an evolving universe.

From the creation narrative in *Genesis* one can keep alive the basic confession that Jhwh calls forth order out of chaos; the creator is seen as the one who causes novelty to arise "out of nothing within the existing state of affairs." His creative word holds out the promise of the emergence of ever-new formations. Plato's narrative, as well, can, with the scientific knowledge we have today, be read in an evolutionary perspective. In this narrative the Demiurge, fixes his gaze at the eternal ideas after which the material world is to be modeled. Yet, in a next move, Plato relativizes the design approach by focusing on the role played by the "World Soul" in bringing about order out of chaos. The "World Soul," carefully constructed by the Demiurge on the basis of mathematical and musical proportions, acts upon the world as the immanently organizing principle; it lures matter into welcoming the imprint of the mathematical and geometrical forms. Thanks to this lure, ever new, unexpected constellations of order come into being.

In Aristotle's cosmology, the highest God is presented as the one who enjoys the fecundity of his thinking and who on this basis exerts a lure on the firmament whose perfect circular rotation is needed for sustaining the processes of growth on earth as well as the transmission of the genetic code in plants and animals from one generation to the other. He is aided in this task by 55 cosmic intelligences (second causes in creation) who in turn lure their planetary sphere towards their perfect rotation. Unlike Plato,

---

28   Robert Asher, *Evolution and Belief,* 231.

Aristotle has no narrative about the beginning of the universe. His God is the one who through a cascade of cosmic mediators —including stars and planets—warrants the durability of life on earth. The continuation of biological life on earth depends on the steady and varied rhythms of the rotating stars and planets. Where does this intuition come from, one might ask, that biological life on earth is linked to what happens in the skies? Equally remarkable is the fact that, in this cosmology, God acts upon the world not through physical intervention but by exerting a lure: "God is the Love who moves the sun and the stars" (Dante).

Finally, in the Christian religious imagination, God calls the cosmos into existence with the help of his Word (*Logos*), a Word that "was with God at the beginning and through whom all things came to be" (Jn. 1: 3). This is a core confession in the Nicene Creed. The fact that the creator acts in union with his logos results in a Creator God who is less external to the world than the God of *Genesis*. True, God the Father creates, in the style of *Genesis*, the world "out of nothing," but, in the Johannine Gospel, we learn that he does so through the mediation of the Son, his creative Word that is going to animate the whole creation from within: as an inner-cosmic principle, comparable to the working of Plato's world soul. This theology of the cosmic Christ later inspired the paleontologist Teilhard de Chardin to develop an evolutionary world view that reaches its apogee in the Omega Point. In the course of this evolution, pre-life (Alpha) emerged through the formation of bio-molecules, followed by the stage of life (from viruses to animals with skeletons, and then mammals) and that of self-consciousness (from pre-hominians to homo sapiens) to eventually attain the stage of human planetisation (with love-energy as the unifying force). The Omega Point of convergence is nothing else but the cosmic Christ, who exerts a potent lure on the whole of the evolutionary process, so that it may attain its full completion.

The Christian religious imagination does not stop at this point. In his reflections on the role of Christ in the bringing forth of creation, Saint Bonaventure develops the vision that all the created entities, according to degrees, give expression to the creative energy that flows from the unfathomable source-deity. They are empowered to activate this energy because they share in the life-form of Christ, the expressive Word that brings the whole abysmal wealth of the Father—of the source-deity—to visible expression. For Bonaventure, it is as if the whole of nature knew its provenance from the abysmal ground of the deity. In their particular mode of existence, rivers, plants, and birds display their own expressive force while, at the same time, giving shape and form to the exuberant creative ground of the deity.

While Bonaventure poetically evokes the astonishing fact of the deity's creative ground that sustains the whole created order, Thomas Aquinas delves into this phenomenon when taking sides in the controversy about the "eternity of the world (the cosmos)." As is well known, Aquinas assimilated Aristotle's world view according to which the whole celestial domain—the celestial spheres, stars and planets, and the 55 auxiliary movers—always existed and will exist for ever, just as the highest God always existed and will exist for ever. The question is then how we have to understand this co-eternity of God and the whole cosmic celestial domain. Aquinas' answer is twofold: he, first of all, states that it is only through revelation that we know that the cosmos has a temporal beginning. Yet, having taken for granted the "eternity of the world," Aquinas then moves on to a philosophical reflection that abstracts the data of revelation from the argument. In this reflection, which is still valid today, he argues that, from the standpoint of the world, it cannot be proven that the celestial entities—especially the 55 auxiliary movers—had a beginning in time. (In a sense he anticipated the world view of scientists like Hawking who hold that the universe—or a proliferation of multiverses—is eternally caught up in a process of self-creation, without any real beginning in time). What can be demonstrated, however, is that those cosmic entities owe their eternal existence to the eternal existence of God, the First Cause and Primordial Origin. There is only one necessary being that *of itself* has the necessity to exist, whom we call God, whereas all the other necessary (that is, eternal) beings receive their necessity to exist *from* this sovereign ground. There is only one God—one source deity in Christian perspective—whose energetic force sustains all the rest.

The above "pre-scientific" approaches to creation have various elements in common. Two of them are of paramount importance: (a) the basic persuasion of ontological dependency on the origin; and (b) the fact that the deity acts upon the entities by exerting a lure upon them. Both aspects are worth considering because they reveal the point at which the mindset of contemporary scientific rationality goes astray.

First, the notion of dependency. Aquinas made it clear that creation basically entails an ontological dependency of the whole of reality on the overwhelming Origin—on the First Cause of all that exists. For religious people, ontological dependency has by no means a negative connotation since it involves the awareness of being empowered by this origin with an energy one is generously allowed to tap and draw upon. A classic of this religious approach is Plotinus. For him, ontological dependency goes hand in hand with participation, according to degrees, in the creative power that overflows from the One. So, to live and to exist is experienced as God-given empowerment, which in turn awakens feelings of marvel and gratitude. To perceive things under the aspect of their reference to the Origin becomes a deeply engrained habit. Not only are the events of one's own life referred to the First Cause, but the working of the whole of nature as well. The special way of thinking in terms of participation and empowerment has its own rationality: a rationality that sets great store by the values of receptivity, and the welcoming of God-given new opportunities. The emphasis on receptivity and gratitude is a fortunate antidote to the rather narrow rationality of contemporary science that seeks primarily to construct an explanatory framework of cause and effect in the logical order of things. This logical investigation can become so rigid that no causal explanation can be regarded as final: things that once functioned as adequate causal explanation suddenly lose this position and are reduced in turn to effects of an anterior chain of causation, etc.—thus leading to an infinite regress in the order of causation. Typical of this explanatory framework is Stephen Hawking's comment on the First Cause: "It is reasonable," he wrote, "to ask who or what created the universe, but if the answer is God, then the question has merely been deflected to that of who created God."[29] No longer is any respect being paid to the metaphysical reality of the Origin.

Second, the primacy of exerting a lure. Unlike the Hebrew God who calls forth order in the world through the sole power of his word, Plato's creation narrative, as a complement to this external command, introduces a world-immanent animating principle, the "World Soul," which lures formless matter towards welcoming the imprint of the eternal ideas. The same idea of exerting a lure figures in Aristotle's account of the manner in which the highest God moves the firmament and the 55 auxiliary movers move their planetary sphere: they all act upon the celestial spheres by exerting a lure on them, that is, they seduce those entities to perform their life-promoting circular rotation. It is upon this model that Whitehead, in *Process and Reality*, described God's action upon the world in terms of lure, invitation, and seduction.[30] This was for him the only way to conceive of a universe endowed with a creativity of its own. If we take this for granted, then we understand better what George Coyne had in mind when stating that "God lets the world be what it will be in its continuous evolution. He is not continually

---

29   Stephen Hawking and Leonard Mlodinow, *The Grand Design*, 172.

30   Alfred North Whitehead, *Process and Reality*, 406 : "He [God] is the lure for feeling, the eternal urge of desire."

intervening, but rather allows, participates, loves."[31] For Coyne, it is a well established scientific fact that the fertility of the universe—through the interplay of chance and necessity—gives directionality to the cosmic evolution. At the same time, he succeeds in giving a theological interpretation to this phenomenon. In terms of a theology of creation, this inherent fertility can be seen now as the universe's response to the lure exerted on it by the seduction of the Creator God. Believers acquainted with ancient cosmologies should have no problems in making this connection: Through his seduction the Creator God empowers the universe with the capacity of a breath-taking productiveness.

The above considerations make it clear that people today, especially scientists who venture into asking ultimate questions about the universe, can learn some lessons from the past. A careful study of the classic sources of creation theology shows that (a) they stress the irreplaceable position of the First Metaphysical Cause of all that exist, and (b) admit of an intrinsic creativity of the cosmos: the One empowers the "finite" (non-divine) entities to share in his marvelous creativity. On this basis, it is possible for believers to regard the world-immanent evolutionary processes as stimulated by the Creator God, an aspect pure science will never be able to convey. Conversely, the achievements of contemporary astrophysics offer a welcome opportunity for deepening one's insight into the God-willed intrinsic creativity of the universe. For without the data provided by them— such as the ripples in the cosmic microwave background radiation, or the twisted light patterns indicative of the gravitational waves that erupted in the awe-inspiring "inflation" of space-time some nanoseconds after the big bang—one would hardly get a concrete picture of the cosmic adventure that led to our stupendous emergence in it.

Times are ripe for theological imagination to conceive of a God who calls forth a universe endowed with an intrinsic creativity in which chance and random outcomes play an important role. The classic theistic God goes hand in hand with a static universe. Once, however, our universe is seen as continuously evolving, that static God must be replaced with a Creator God who is understood to be so generous that he allows the creation to share in his own creativity.

---

31   George Coyne, "God's Chance Creation," *The Tablet,* August 6, 2005.

Index

Father George, in a photograph with his doctoral student, Father Dave Gentry-Akin, who shepherded the US publication of this book after Father Georges' untimely death during a teaching stint in the Philippines.

Father Georges DeSchrijver, SJ
Born March 15, 1935
Died October 7, 2016

"And those who are wise shall shine like the brightness of the firmament, and they that instruct many to justice, as stars for all eternity" (Dan 12:3).